OPERATIONS PLANNING

Mixed Integer Optimization Models

The Operations Research Series

Series Editor: A. Ravi Ravindran

Professor, Department of Industrial and Manufacturing Engineering
The Pennsylvania State University – University Park, PA

Published Titles:

Operations Planning: Mixed Integer Optimization Models
Joseph Geunes

Introduction to Linear Optimization and Extensions with MATLAB®
Roy H. Kwon

Supply Chain Engineering: Models and Applications
A. Ravi Ravindran & Donald Paul Warsing

Analysis of Queues: Methods and Applications
Natarajan Gautam

Integer Programming: Theory and Practice
John K. Karlof

Operations Research and Management Science Handbook
A. Ravi Ravindran

Operations Research Applications
A. Ravi Ravindran

Operations Research: A Practical Introduction
Michael W. Carter & Camille C. Price

Operations Research Calculations Handbook, Second Edition
Dennis Blumenfeld

Operations Research Methodologies
A. Ravi Ravindran

Probability Models in Operations Research
C. Richard Cassady & Joel A. Nachlas

OPERATIONS PLANNING

Mixed Integer Optimization Models

Joseph Geunes

CRC Press is an imprint of the
Taylor & Francis Group, an **informa** business

CRC Press
Taylor & Francis Group
6000 Broken Sound Parkway NW, Suite 300
Boca Raton, FL 33487-2742

First issued in paperback 2017

Version Date: 20140805

ISBN 13: 978-1-4822-3990-4 (hbk)
ISBN 13: 978-1-138-07478-1 (pbk)

Visit the Taylor & Francis Web site at
http://www.taylorandfrancis.com

and the CRC Press Web site at
http://www.crcpress.com

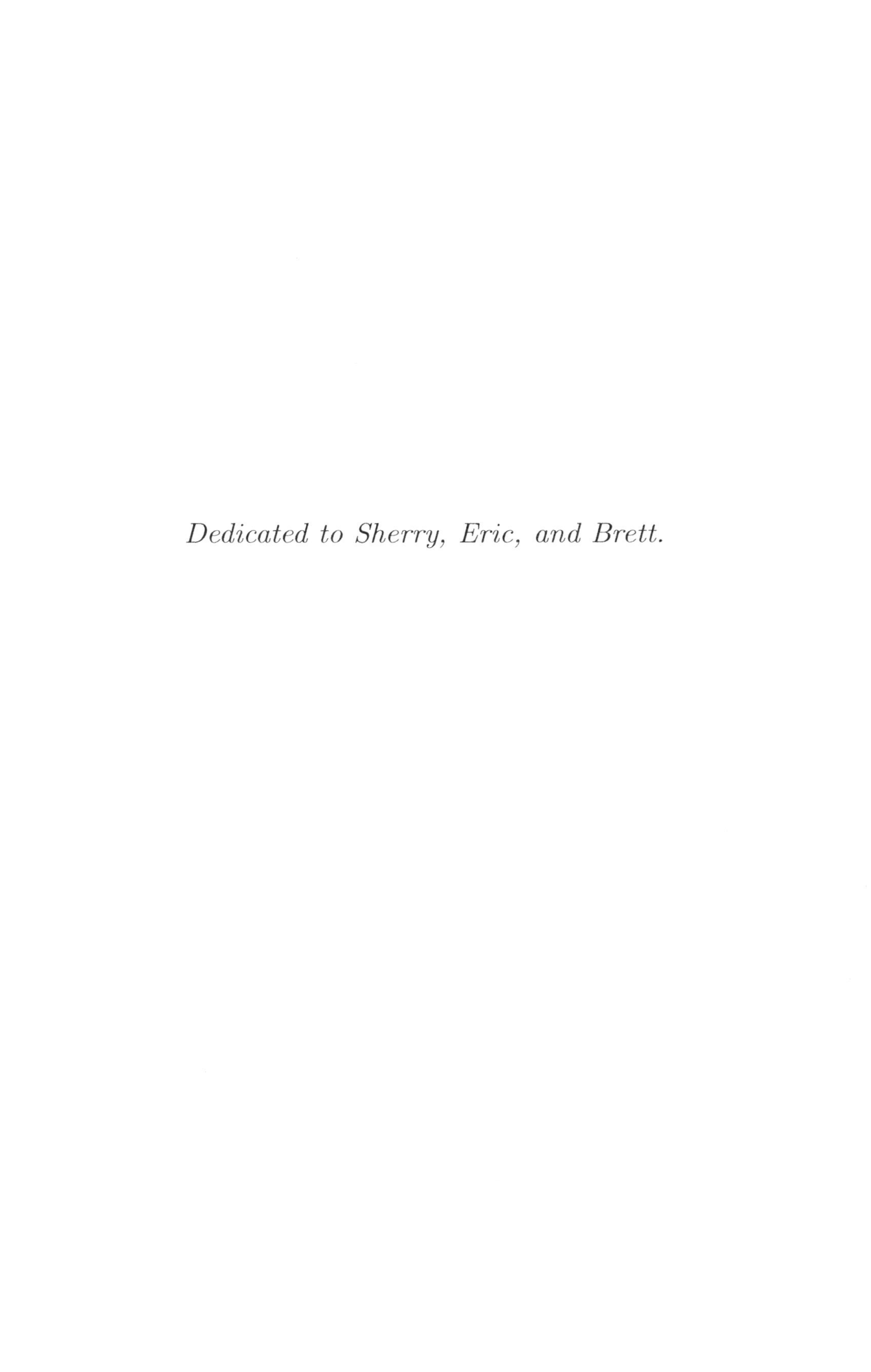

Dedicated to Sherry, Eric, and Brett.

Contents

Preface **xi**

Author Biography **xiii**

1 Introduction and Purpose **1**

1.1 Operations Planning 1
1.2 Mixed Integer Optimization 2
1.3 Optimization Models in Operations Planning 4

2 The Knapsack Problem **7**

2.1 Introduction 7
2.2 Knapsack Problem 0-1 Programming Formulation 8
2.2.1 Relation to the subset sum problem 11
2.3 Linear Relaxation of the 0-1 Knapsack Problem 11
2.4 Asymptotically Optimal Heuristic 15
2.5 Fast Approximation Algorithm 17
2.6 Valid Inequalities 19
2.7 Review 23

3 Set Covering, Packing, and Partitioning Problems **25**

3.1 Introduction 25
3.2 Problem Definition and Formulation 26
3.3 Solution Methods 27
3.3.1 Bin packing heuristics 28
3.3.2 Column generation and the set partitioning problem . 29
3.3.3 Branch-and-price for the set partitioning problem . . . 33
3.4 Review 37

4 The Generalized Assignment Problem **39**

4.1 Introduction 39
4.2 GAP Problem Definition and Formulation 40
4.3 Lagrangian Relaxation Technique 41
4.3.1 Lagrangian relaxation for the GAP 46

4.4 Branch-and-Price for the GAP 47
4.5 Greedy Algorithms and Asymptotic Optimality 49
4.6 Review 54

5 Uncapacitated Economic Lot Sizing Problems 57

5.1 Introduction 57
5.2 The Basic UELSP Model 58
5.2.1 Fixed-charge network flow interpretation 59
5.2.2 Dynamic programming solution method 61
5.3 Tight Reformulation of UELSP 62
5.3.1 Lagrangian relaxation shows a tight formulation . . . 63
5.4 An $\mathcal{O}(T \log T)$ Algorithm for the UELSP 65
5.5 Implications of Backordering 69
5.6 Review 71

6 Capacitated Lot Sizing Problems 73

6.1 Introduction 73
6.2 Capacitated Lot Sizing Formulation 74
6.3 Relation to the 0-1 Knapsack Problem 75
6.3.1 Fixed-charge network flow interpretation 76
6.3.2 Dynamic programming approach 77
6.4 The Equal-Capacity Case 78
6.5 FPTAS for Capacitated Lot Sizing 80
6.5.1 Structure of the dynamic programming approach . . . 80
6.5.2 Approximation of the dynamic program 83
6.6 Valid Inequalities for the CELSP 85
6.6.1 (S, l) inequalities 86
6.6.2 Facets for the equal-capacity CELSP 87
6.6.3 Generalized flow-cover inequalities 90
6.7 Review 93

7 Multistage Production and Distribution Planning Problems 95

7.1 Introduction 95
7.2 Models with Dynamic Demand 97
7.2.1 Serial systems with dynamic demand 97
7.2.2 Production networks with non-speculative costs 101
7.2.3 Constant-factor approximations for special cases . . . 105
7.3 Models with Constant Demand Rates 112
7.3.1 Stationary, nested, power-of-two policies 114
7.3.2 The joint replenishment problem 128
7.3.3 The one-warehouse multi-retailer problem 130
7.4 Review 134

8 Discrete Facility Location Problems 137

8.1 Introduction . . . 137
8.2 Relation to Previous Models in this Book . . . 138
8.2.1 Cost-minimizing version of the FLP . . . 141
8.2.2 Relationship of the FLP to lot sizing problems . . . 141
8.2.3 Single-sourcing version of the FLP and the GAP . . . 142
8.2.4 Set covering and FLP complexity . . . 142
8.3 Dual-Ascent Method for the Uncapacitated FLP . . . 143
8.4 Approximation Algorithms for the Metric UFLP . . . 148
8.4.1 Randomization and derandomization . . . 151
8.5 Solution Methods for the General FLP . . . 154
8.5.1 Lagrangian relaxation for the FLP . . . 154
8.5.2 Valid inequalities for the FLP . . . 157
8.5.3 Approximation algorithms for the FLP . . . 162
8.6 Review . . . 165

9 Vehicle Routing and Traveling Salesman Problems 167

9.1 Introduction . . . 167
9.2 The TSP Graph and Complexity . . . 168
9.3 Formulating the TSP as an Optimization Problem . . . 170
9.4 Comb Inequalities . . . 173
9.5 Heuristic Solutions for the TSP . . . 175
9.5.1 Nearest neighbor heuristic . . . 175
9.5.2 The sweep method . . . 177
9.5.3 Minimum spanning tree based methods . . . 179
9.5.4 Local improvement methods . . . 182
9.6 The Vehicle Routing Problem . . . 182
9.6.1 Exact solution of the VRP via branch-and-price . . . 183
9.6.2 A GAP-based heuristic solution approach for the VRP 185
9.6.3 The Clarke-Wright savings heuristic method . . . 187
9.6.4 Additional heuristic methods for the VRP . . . 190
9.7 Review . . . 191

Bibliography 193

Index 201

Preface

This book is intended for researchers and graduate students working at the intersection of deterministic operations research and operations planning. It may serve as a desk reference for researchers or as a textbook for a graduate course on mixed integer optimization models for operations planning. Each chapter considers a particular classical operations planning problem class, and examples are used throughout to demonstrate the problem-solving methods described; the end of each chapter (with the exception of Chapter 1) contains a review of the chapter and a set of exercises. While each chapter presents a given problem class, the various problem classes are closely related to one another in interesting ways that are not obvious at first blush. The connections between the various models covered, therefore, serve as one of the central themes throughout this book.

Defining the scope and coverage of topics in this work proved to be something of a challenge because of the sheer volume of work that might arguably fit with the book's title. And, while the real world presents us with many sources of uncertainty that confound our ability to plan operations ahead of time, deterministic models, nevertheless, continue to play an important role in operations planning. In particular, practical operations planning contexts exist such that future customer needs may be treated as known ahead of time, as is often the case in make-to-order operations. Additionally, future costs and operations planning constraints are often either known in advance or may be estimated with a high degree of precision in the planning stages. As a result, the models and methods presented throughout this book find widespread application in practice.

As a former graduate student, as well as an advisor to graduate students in the areas of operations research and operations planning for more than fifteen years, I learned about the models and problems covered in this book (and their interrelationships) in a somewhat piecemeal fashion, from numerous, disparate sources. This book is, therefore, intended to serve as a more direct route to gaining a fundamental understanding of several of the critical and classical problem classes that regularly appear in the problems we address in our field.

Each of the models and solution methods presented in this book is the result of high-impact research that has been published in the scholarly literature, and appropriate references are cited throughout the book. While these past works have been essential to my own work, credit must be given to the past researchers whose work is presented herein, and I am eternally grateful to those who have made the invaluable contributions that follow. I apologize in advance for any errors and/or omissions, which are my responsibility alone.

I would like to express my gratitude to Professor A. Ravi Ravindran, the editor of this book series, who provided valuable advice in the planning stages of this book. In addition, I would like to thank Melis Teksan for graciously offering her proofreading and editorial talents. Finally, I greatly appreciate the seamless and efficient manner in which Cindy Carelli of CRC Press managed the development process, as well as the assistance provided by Kate Gallo of CRC Press in seeing this project through to completion.

Joseph Geunes
Gainesville, Florida

Author Biography

Joseph Geunes received his Ph.D. in 1999 from the Pennsylvania State University in Operations Research and Business Administration. Prior to this he received an MBA from the Pennsylvania State University in 1993 and a B.S. in Electrical Engineering from Drexel University in 1990. Since 1998, he has been a faculty member in the Department of Industrial and Systems Engineering at the University of Florida, where he is currently Professor and Chair. His research focuses on applied operations research in production, inventory, and supply chain management. Professor Geunes is a Department Editor for *IIE Transactions* and an Area Editor for *Omega*, and has advised or co-advised 15 students who have completed their Ph.D. degrees.

1

Introduction and Purpose

1.1 Operations Planning

Within the broad field of business administration, the area of business operations encompasses the activities required to produce goods and services and bring them to markets and customers. Executing the associated production and distribution activities requires material inputs as well as physical and organizational infrastructure. Physical infrastructure is comprised of the facilities (e.g., factories, warehouses, retail stores) and equipment (e.g., production and material handling equipment, inter-facility transportation vehicles) necessary to transform material inputs into production outputs that provide value to customers. Organizational infrastructure consists of the human resources required to ensure the execution of operations activities, as well as the functional responsibilities and relationships among the organization's members.

Execution of operations activities requires *planning* in addition to infrastructure and materials. At some point, the establishment of every existing factory, warehouse, and retail outlet was planned. This plan began with the identification of the need for the facility, its defined mission, and a characterization of the scope of interaction with other facilities and organizations required to accomplish the planned mission. Similarly, the broad responsibilities and functions of each organizational member were, at some point, planned and defined. Such infrastructure and organizational plans often require initial investments and are put in place for a relatively long time horizon (e.g., five years or more). These long-range planning activities fall within the domain of *strategic planning*, and they serve as a basis for achieving an organization's set of long-term, strategic goals (e.g., to provide a particular class of products or services to a particular set of markets).

After establishing the infrastructure required to implement a strategy, the medium-range scope of activities performed at facilities and the interactions between facilities and their outputs require planning as well. Such plans include establishing targeted quarterly workforce levels, required material inputs, production outputs, and material and finished goods inventory levels at each facility over a one- or two-year horizon, for example. These medium-range plans are critical for determining operating capital requirements, contractual arrangements for material supplies and distribution services, and projected earnings for a firm, for example. Planning for these medium-range activities is often referred to as *tactical planning.*

Reaching markets with products and services requires daily operations activities, including the production of actual goods and the movement of goods between locations. On any given day (or in any given week or month), for example, the workers in a facility need to process a set of activities in order to meet some desired output by the end of the day (or the end of the week or month). That is, given a set of requirements for production output, a sequence in which these requirements will be processed must be determined. Such work sequencing decisions typically impact the economic performance of a facility as well as the facility's ability to meet customer expectations. These short-range work sequencing decisions fall within the domain of *operations planning.*

Each of these planning categories falls within an overall planning *hierarchy*, wherein the decisions made at higher levels of the hierarchy constrain the available decisions at lower levels. Strategic planning serves as the top level in this hierarchy, tactical planning as the middle level, and operations planning as the lowest level. The intent of this conceptual hierarchy is to decompose the entire scope of planning activities into a set of manageable decisions, because the amount of information required and the complexity involved often preclude simultaneously considering decisions at each level of the hierarchy. In order to mitigate the possible negative impacts of decomposing decisions into such a hierarchy, decisions at higher levels often use rough-cut or aggregated estimates of the required activity levels at lower levels in the hierarchy. For example, when planning the construction of a facility (a strategic planning decision), the facility's capacity level may be determined based on the projected maximum annual demand the facility is intended to satisfy. This projected maximum annual demand is a function of the aggregate anticipated requirements over a medium-range planning horizon (e.g., one year), which are, in turn, aggregated estimates of the requirements during the short-term (e.g., weekly). Planning at each level of the hierarchy requires a systematic, methodological approach in order to ensure that the plans are consistent at all levels and allow an organization to effectively execute its strategy.

1.2 Mixed Integer Optimization

Problems arising over the last century in the production and distribution of goods and services have led to numerous mathematical modeling and scientific approaches for operations planning decisions. Development of these approaches was accelerated following the conception of Linear Programming (LP) by George B. Dantzig in 1947 (see [36] for a history of the advent of LP as told by Dantzig himself) and the ensuing development of the field of operations research. The broader field of mathematical programming generalizes the fundamental concepts of LP and has provided a powerful paradigm for the mathematical representation of complex decision problems.

Each of the operations planning problems discussed in this book is cast as a special kind of mathematical program known as a mixed integer optimization problem. Thus, we require an ability to express each of the problems we consider in terms of a mathematical model. Mathematical programs are characterized by a quantifiable set of *decision variables* that often translate to the level at which an activity will be performed. Examples include the number of units of production in each time period at a production facility, the number of units transported from a facility to a warehouse in a given time period, and the collection of jobs assigned to a machine for production in a given day. In addition to decision variables, mathematical programs incorporate a set of *constraints* that define the allowable values for the decision variables. Examples of constraints include a limit on the number of units of product that can be transported on a vehicle on a given day or a limit on the number of jobs a machine may process in a day. The final necessary element for defining a mathematical program is a so-called *objective function*, which consists of a mathematical expression stated in terms of the decision variables, which the decision maker wishes to minimize (e.g., cost) or maximize (e.g., profit).

Next, providing a more precise definition of the general class of mathematical programs in a form that is easy to communicate requires introducing some basic notation. Let $x \in \mathbb{R}^n$ denote an n-dimensional vector of decision variables with elements $[x_1, x_2, \ldots, x_n]$ (where $\mathbb{R}^n$ denotes the set of all real vectors in n-space), and define $f(x) : \mathbb{R}^n \rightarrow \mathbb{R}$ as a function we would like to minimize (e.g., $f(x)$ gives the cost associated with the vector x). Let $X \subset \mathbb{R}^n$ denote the set of *feasible* solutions, i.e., the set of n-dimensional vectors in $\mathbb{R}^n$ that satisfy a prescribed set of constraints on the allowable decision variable values. Then we can express a generic mathematical program as follows:

[MP]

$$\text{Minimize} \quad f(x) \tag{1.1}$$

$$\text{Subject to:} \quad x \in X. \tag{1.2}$$

The class of linear programs consists of problems in the form of MP such that $f(x)$ is a linear function and X corresponds to a set of linear inequalities, including the condition $x \geq 0$ in standard form. The class of mixed integer linear programming (MILP) problems requires the added condition that some subset of the variables must take integer values.

Not only do the majority of the problems covered in this book fall within the category of MILP problems, but they also fall within a subset of this class in which some of the integer variables must take a value of zero or one, i.e., a subset of the integer variables must be binary. While this may, at first glance, appear to be a limiting factor, as we will see, numerous widely used and broadly applicable operations planning problems fall within this class. This is because binary decision variables provide a powerful construct for characterizing logical constraints, and permit quantifying relationships implied by logical conditions. That is, we can think of a binary variable as an on-off condition or a "switch." For example, suppose we are considering whether or not to build a facility that will have a capacity to produce K units of a good per year, and let δ denote a binary variable equal to one if the facility is built, and equal to zero otherwise. Let y denote the number of units that will be produced at the facility per year. Then the possible values of y are limited by the condition $0 \leq y \leq \delta K$. If $\delta = 1$ then the facility is built, and y must take a value between zero and K. On the other hand, if the facility is not built, then the only possible value of y is zero. This use of binary variables as "yes-no" or "on-off" switches occurs throughout the models presented in this book, as this facilitates expressing various operations planning problems as MILPs.

1.3 Optimization Models in Operations Planning

Starting with the knapsack problem in Chapter 2, this book presents a set of classical optimization models with widespread applications in operations planning. The discussion of each classical model begins with a motivation for studying the problem, and includes examples of the problem's application in operations planning contexts. In addition to providing a mathematical problem statement in the form of a mathematical program, for each problem class, we explore special structural results and properties of optimal solutions that lead to effective algorithmic solution approaches for the problem class.

Thus, as an important by-product of studying mixed integer optimization models for operations planning, we cover the essentials of several important techniques for modeling and solving mixed integer optimization problems in general. In particular, we will cover the following techniques and concepts:

- Computational Complexity: Chapter 2 provides a brief overview of $\mathcal{NP}$-Completeness and the class $\mathcal{P}$ of polynomially solvable problems;
- Asymptotically Optimal Heuristics: In studying the knapsack problem and generalized assignment problem in Chapters 2 and 4, respectively, we consider asymptotically optimal heuristics under a suitable probabilistic model for the problem's data. These heuristic approaches guarantee that, as the problem size grows, the percentage deviation from optimality of the resulting solution approaches zero;
- Fully Polynomial Time Approximation Schemes (FPTAS): When studying the knapsack and lot sizing problems in Chapters 2 and 6, we describe the development of this powerful method for efficiently finding near-optimal solutions for difficult problems, with performance bounds;
- Constant-Factor Approximations: When a heuristic method guarantees a solution that is no more than a fixed constant multiplied by the optimal

solution value, regardless of the problem's size, this is known as a constant-factor approximation algorithm. Chapters 7 and 8 cover constant-factor approximation algorithms for multistage production and distribution planning problems and metric facility location problems, respectively;

- Branch-and-Price: The development of solution methods for set partitioning, generalized assignment, facility location, and vehicle routing problems (Chapters 3, 4, 8, and 9, respectively) incorporates column-generation methods for solving large-scale linear programming relaxations and using these relaxation solutions within a branch-and-bound algorithm;
- Lagrangian Relaxation: Chapter 4 contains a basic primer on Lagrangian relaxation for difficult mixed integer optimization problems with complicating constraints. This method is very useful for solving multi-resource problems such as the generalized assignment problem (Chapter 4), multistage production and distribution planning problems (Chapter 7), and capacitated facility location problems (Chapter 8);
- Valid Inequalities: While this book does not emphasize polyhedral theory or results, we do introduce several intuitive valid inequalities that are useful for strengthening the linear programming relaxation bound for difficult mixed integer programs.

An area of particular emphasis in this book lies in drawing out and highlighting the close relationships among the various models considered. When studying the models covered within this book, students often learn about them sequentially and in isolation from each other, and the inherently close relationships are not always obvious when approached in this way. One of the primary goals of this book is, therefore, to clearly highlight those situations in which a particular model results as a special case of other related models or how one model generalizes another. Understanding these relationships allows a researcher to more easily characterize new models, through their relationships to the classical models covered in this book.

The audience for this book, therefore, consists of graduate students and researchers working at the interface between operations research and operations planning. For graduate students, this book may serve as a text for a course on fundamental mixed integer optimization models for operations planning. For researchers and practitioners, this book is intended to serve as a useful reference on models that find widespread applications in operations. In addition, by considering these models together and exploring their relationships with one another, the goal is to assimilate various critical operations research techniques and classical operations planning models without the need to consult numerous disparate sources throughout the academic literature.

Each of the chapters that follows covers a particular classical problem class, which has served as the subject of intensive research for decades. The modeling and solution methods presented within each chapter are attributable to the authors of the various papers cited throughout.

2

The Knapsack Problem

2.1 Introduction

The knapsack problem appears as a fundamental subproblem in seemingly countless optimization problems, and is a deceptively simple looking problem that can be used to illustrate the attention to detail required in analyzing optimization models. The problem can be easily explained as follows. Suppose you have a knapsack with a capacity that can be measured using a single dimension (e.g., weight). You have a number of individual items that you would like to take with you in the knapsack, but the collective set of items will not fit in the knapsack. That is, the collective capacity consumption of the individual items exceeds the knapsack's capacity. Each of these items has a value to you and you would like to maximize the total value of the items that you take with you in the knapsack. The goal is to determine the subset of items to put in the knapsack (without exceeding the knapsack capacity) that maximizes the total value of items you carry with you. Clearly if each item consumes exactly one unit of knapsack capacity, then the problem is trivial. We simply need to sort the items in nonincreasing value order (breaking ties arbitrarily), and insert them into the knapsack in this order until either the knapsack is full, or until we have inserted all items with positive (or nonnegative) value. Similarly, if all items have the same value, then we insert items into the knapsack in nondecreasing order of capacity consumption levels. When the items consume different levels of capacity and have different values, however, the problem becomes substantially more difficult.

Although motivated by this "knapsack" example, this problem can be viewed as a more generic resource allocation problem, where the knapsack capacity corresponds to some measure of capacity of a generic resource, and the items correspond to activities that consume resource capacity. For example, a machine may have a certain number of operating minutes per day, and each member of a set of jobs that we would like to perform on the machine requires a certain number of minutes of processing time. Or, a truck can carry a certain total weight, and we would like to ship a number of items on the truck, but the sum of their weights exceeds the truck's weight capacity. Generic resource capacity constraints of this type appear as a subset of a seemingly endless number of optimization problems in practice, which has made the knapsack problem one of the most well studied and understood optimization problems.

The knapsack problem, like several of the problems we will consider in this book, serves as the subject of a number of full-length texts (see, for example, Martello and Toth [84] and Kellerer, Pferschy, and Pisinger [69]) that explore numerous variants of the fundamental problem. In this chapter, we study the special structure of the one-dimensional knapsack problem with indivisible items, and highlight effective solution methods for this problem. Understanding this problem is fundamental for operations modeling, and we use this context to provide examples of techniques in asymptotic heuristic analysis, fully polynomial time approximation schemes, and the use of valid inequalities to strengthen the linear programming relaxation of a problem with integer variable requirements. To explore multiple-dimension and multiple-knapsack generalizations of this problem, as well as advanced computational and implementation issues, the reader is encouraged to refer to the aforementioned textbooks, as well as Chapter 4 on the generalized assignment problem (GAP).

2.2 Knapsack Problem 0-1 Programming Formulation

We begin by defining the knapsack problem data, which include the knapsack capacity K and the set I of n items, $I = \{1, 2, \ldots, n\}$. For item $i \in I$ we let c_i and b_i denote the value and capacity consumption of the item, respectively. We assume without loss of generality that the parameters K and b_i for all $i \in I$ take integer values, while each c_i is a real number. Because each item requires a *yes* or *no* decision on whether we will insert it into the knapsack, we create a binary variable x_i for each $i \in I$, which equals one if we insert item i into the knapsack, and zero otherwise. The classical 0-1 knapsack problem can then be represented by the following 0-1 programming formulation.

[KP] Maximize
$$\sum_{i \in I} c_i x_i \tag{2.1}$$
Subject to:
$$\sum_{i \in I} b_i x_i \leq K, \tag{2.2}$$
$$x_i \in \{0, 1\}, \quad i \in I. \tag{2.3}$$

We assume that a feasible solution exists for the KP, i.e., that some subset $S \subset I$ exists such that $\sum_{i \in S} b_i \leq K$. At this point we have said nothing about the sign of the objective function and constraint coefficients, or about the sign of the capacity limit K. We next show that we are free to assume, without loss of generality, that all problem data are positive in the KP. Suppose we have a problem where this is not the case, and let $S_n \subset I$ denote the set of all $i \in I$ such that $b_i < 0$. Suppose that for each $i \in S_n$ we replace the associated x_i by its complement, $x_i = 1 - x_i'$ (where $x_i' \in \{0, 1\}$), which leads to an increase in the right-hand side by $-b_i$. Then the resulting equivalent problem has positive values for all b_i. Next suppose that the original value of

K was negative. Problem feasibility requires $\sum_{i \in S_n} b_i \leq K$, and after taking the complement ($x_i = 1 - x'_i$) for each item in S_n, the new capacity value (which equals $K - \sum_{i \in S_n} b_i \geq 0$) must now be nonnegative. We can therefore assume without loss of generality that all b_i values and K are positive (we do not consider $K = 0$ because the only resulting feasible solution sets all $x_i = 0$; for any $b_i = 0$ we can fix x_i based on the sign of the objective function coefficient c_i, independent of the constraint). We can further assume without loss of generality that $K \geq \max_{i \in I}\{b_i\}$ since feasibility requires setting $x_i = 0$ for any item such that $b_i > K$. After making substitutions required to ensure positive values for all b_i and K, we can assume without loss of optimality that $c_i > 0$ for all i, since any item violating this condition will have $x_i = 0$ in an optimal solution. Note that this implies that any original item such that $c_i > 0$ and $b_i < 0$ will have $x_i = 1$ in an optimal solution, which satisfies intuition because such items serve to simultaneously add to the objective function and increase the available capacity. Finally, we only consider cases such that $\sum_{i \in I} b_i > K$ to avoid trivial problem instances where the constraint is irrelevant to the problem.

An exact solution for the 0-1 knapsack problem can be obtained using dynamic programming. Although this problem falls into the class of $\mathcal{NP}$-Hard optimization problems,[1] this dynamic programming solution can often be obtained in reasonable computing time for practical problem sizes. The solution time is a function of the product of the number of items n and the knapsack capacity K. To see this, we next provide a dynamic programming formulation for the 0-1 knapsack problem. Define $f_i(k)$ as the optimal solution value of a knapsack problem with capacity k that includes items 0 through i, where item 0 denotes a 'dummy' item that consumes zero capacity and has zero value. Clearly $f_i(0) = 0$ for all $i = 1, 2, \ldots, n$ and $f_0(k) = 0$ for all $0 \leq k \leq K$, and these serve as a set of boundary conditions for the dynamic programming solution. For any i and k such that $b_i \leq k$, we can compute $f_i(k)$ by comparing the solution in which we do not insert item i into the knapsack to the solution where item i is inserted into the knapsack. That is, $f_i(k)$ can be computed recursively beginning with the boundary conditions as

$$f_i(k) = \max\left\{f_{i-1}(k), c_i + f_{i-1}(k - b_i)\right\}, \tag{2.4}$$

where the first term in the maximization represents the maximum value if we do not insert item i into the knapsack of size k (including items $1, \ldots, i-1$ only), and the second term provides the maximum value if we do insert item i. If we define $f_i(k) = -\infty$ for any $k < 0$, then we can, in the worst case, begin with $k = 1$ and successively compute (2.4) for each value of i from 1 to n. We then increment k by 1 and continue this process until $k = K$ and

[1] We will have more to say about characterizing the complexity of optimization problems. For our purposes now, it is sufficient to state that the $\mathcal{NP}$-Hardness of the knapsack problem implies that no known algorithm exists that can guarantee solving the problem to optimality in a number of computational steps (or, equivalently, time) that is bounded by a polynomial function of the number of items, n.

$i = n$, with an optimal solution value given by $f_n(K)$. The total solution time is then equal to some constant multiplied by the product of K and n, and we say that the computational complexity of this approach is bounded by a number that is *on the order of* the product of K and n, which we write as $\mathcal{O}(nK)$.

In stating this complexity bound, observe that the parameter n is a measure of the number of elements of input data, while K is a measure of the size of one of the individual data elements. When a solution algorithm has a computational-time performance bound that is a polynomial function of the number of elements of input data only, then the result is a polynomial time solution method, with a solution time that grows as a polynomial function of the *problem size*. Problems that have *polynomial* time solution algorithms fall in the class of problems known as $\mathcal{P}$ in the language of computational complexity theory (please see Garey and Johnson [54] for a comprehensive discussion of computational complexity theory). When a solution algorithm has a performance bound that is a polynomial function of the problem size *and* the magnitude of one or more of the data values, then we have a *pseudopolynomial* time algorithm, which is the best we can hope for in the case of an $\mathcal{NP}$-Hard optimization problem, unless $\mathcal{P} = \mathcal{NP}$.[2] The pseudopolynomial time dynamic programming approach for the 0-1 knapsack problem certainly works very well for reasonable values of knapsack capacity K, although it can become intensive for very large values of capacity. For this reason and for convenient implementation, branch-and-bound methods are often used to solve the 0-1 programming formulation of the knapsack problem using commercial mixed integer linear programming (MILP) solvers. A branch-and-bound algorithm typically requires solving the linear relaxation of the 0-1 programming formulation as a subroutine. It is therefore useful to study efficient solution methods for this linear relaxation, for use within a branch-and-bound algorithm.[3] This linear relaxation also has a practical interpretation in its own

[2]Problems in the class $\mathcal{NP}$ can be solved in polynomial time using the concept of a *nondeterministic Turing machine*, which is a theoretical construct and does not exist in reality (see Garey and Johnson [54]). Problems in the class $\mathcal{NP}$ are posed such that they have a yes/no answer (for example, given a set of items each with a corresponding integer-valued weight, does a subset exist with weights that add up to some integer number C?). Given a solution that satisfies the "yes" condition, the "yes" answer can be verified in polynomial time for problems in the class $\mathcal{NP}$. An $\mathcal{NP}$-Complete problem is one in $\mathcal{NP}$ such that any other problem in $\mathcal{NP}$ can be reduced to that problem in polynomial time. An $\mathcal{NP}$-Hard problem is at least as hard as an $\mathcal{NP}$-Complete problem, and may be one that does not have a yes/no answer or whose "yes" answer cannot be verified in polynomial time (thus an $\mathcal{NP}$-Hard problem may not even be in the class $\mathcal{NP}$). For example, an optimization problem does not strictly speaking have a yes/no answer, but can be solved by a series of problems that do. An algorithm that can solve an $\mathcal{NP}$-Hard problem can then be translated into one that can solve any other problem in $\mathcal{NP}$. Thus if one could provide a polynomial time algorithm that solves an $\mathcal{NP}$-Hard problem, then this algorithm could solve any $\mathcal{NP}$-Complete problem in polynomial time, which would imply $\mathcal{P} = \mathcal{NP}$, although no such known algorithm currently exists.

[3]For details on branch-and-bound algorithm strategies for the knapsack problem, please refer to Salkin and Mathur [103].

right, and provides a rich context for understanding fundamental linear programming concepts. We will explore this linear relaxation in more detail in Section 2.3.

2.2.1 Relation to the subset sum problem

The subset sum problem is an $\mathcal{NP}$-Complete problem (see [54]) that is also a special case of the 0-1 knapsack problem, and may be posed very simply as follows: Given a set $\mathcal{S}$ of integers $\{s_1, s_2, \ldots, s_n\}$, does a subset $\mathcal{I} \subseteq \mathcal{S}$ exist such that $\sum_{i \in \mathcal{I}} s_i = S$ for some integer S? Clearly this question has a yes/no answer for any problem instance, and given a problem instance and a corresponding solution that produces a "yes" answer, its validity can be demonstrated in polynomial time; the subset sum problem thus falls in the class $\mathcal{NP}$. Observe that the 0-1 knapsack problem can be approached by solving a sequence of problems with yes/no answers. For example, using our definition of the 0-1 knapsack problem, suppose we ask the question of whether a subset $\mathcal{I} \subseteq I$ exists such that $\sum_{i \in \mathcal{I}} b_i \leq K$ and $\sum_{i \in \mathcal{I}} c_i \geq P$ for some integer P. If the answer is yes, then we have a solution with a corresponding lower bound on the value of the optimal 0-1 knapsack problem objective. If the answer is no, then we know that the optimal solution is less than or equal to P. It is then possible to perform a binary search on the value of P in order to determine the optimal objective function value for the 0-1 knapsack problem. Thus, if we can answer our yes/no question in polynomial time, it is possible to solve the optimization problem in polynomial time as well. To show that the subset sum problem is a special case of the yes/no version of the 0-1 knapsack problem, consider a special case of the 0-1 knapsack problem such that $b_i = c_i = s_i$ for each $i \in I$ and $K = P = S$. Then our yes/no version of the 0-1 knapsack problem asks whether a subset $\mathcal{I} \subseteq I$ exists such that $\sum_{i \in \mathcal{I}} s_i \leq S$ and $\sum_{i \in \mathcal{I}} s_i \geq S$, which is equivalent to the condition $\sum_{i \in \mathcal{I}} s_i = S$. We will return to the subset sum special case of the knapsack problem in Chapter 6 when discussing the relationship between the knapsack problem and the capacity-constrained version of the economic lot sizing problem.

2.3 Linear Relaxation of the 0-1 Knapsack Problem

The linear programming (LP) relaxation of the knapsack problem is obtained by simply allowing the x_i variables to take any real values between 0 and 1, which results in the following formulation.

$$\textbf{[LP1]} \qquad \text{Maximize} \quad \sum_{i \in I} c_i x_i \tag{2.5}$$

$$\text{Subject to:} \quad \sum_{i \in I} b_i x_i \leq K, \tag{2.6}$$

$$0 \leq x_i \leq 1, \quad i \in I. \tag{2.7}$$

This continuous knapsack problem has applications in its own right, for example, for problems in which items are infinitely divisible. This formulation is equivalent to one in which we make the variable substitution $y_i = b_i x_i$ for all $i \in I$, which can be written as

$$\textbf{[LP2]} \qquad \text{Maximize} \quad \sum_{i \in I} (c_i / b_i) y_i \tag{2.8}$$

$$\text{Subject to:} \quad \sum_{i \in I} y_i \leq K, \tag{2.9}$$

$$0 \leq y_i \leq b_i, \quad i \in I. \tag{2.10}$$

Formulation LP2 serves to clearly illustrate why a *greedy algorithm* provides an optimal solution for this problem. By considering this formulation, it should be clear that items with higher values of the ratio c_i/b_i are preferred. This ratio is intuitively appealing, as we can think of it as a measure of the "bang for the buck" of an item, effectively capturing the value an item provides per unit of capacity consumption. Since each unit of an item in the y-variable space consumes the same amount of capacity, we should use each unit of capacity to obtain the highest value possible, where the value is now measured by the ratio c_i/b_i. We should therefore take the item with the highest value of c_i/b_i and insert as much of it into the knapsack as possible. We stop increasing the value of y_i when it equals the minimum between b_i and the remaining available capacity. If any available capacity remains, we consider the item with the next highest value of the ratio c_i/b_i. We repeat this, breaking ties arbitrarily, until the knapsack capacity is filled (or until we run out of items), and terminate with an optimal solution to the linear program. Algorithm 2.1 provides a sketch of this solution approach.

Algorithm 2.1 Knapsack linear relaxation greedy algorithm.

1: Sort items in nonincreasing order of c_i/b_i, set counter $l = 1$, $\hat{K} = K$, and $y_i = 0$ for $i = 1, 2, \ldots, n$.
2: If $b_l > \hat{K}$, then set $y_l = \hat{K}$ and stop with optimal solution where $y_i = b_i$ for $i = 1, 2, \ldots, l-1$, $y_l = \hat{K}$, and $y_i = 0$ for $i = l+1, \ldots, n$. Otherwise continue.
3: Set $y_l = b_l$, $l = l + 1$, $\hat{K} = \hat{K} - b_l$, and return to Step 2.

To show that this approach provides an optimal solution, we consider the optimality conditions for the linear program LP2. For LP2, given a basic feasible solution,[4] the *reduced cost* of a variable x_i can be written as $c_i/b_i - c_B \mathrm{B}^{-1} A^i$, where c_B denotes the vector of objective function coefficients associated with the basic variables, B denotes the basis matrix (i.e., the matrix consisting of the columns associated with the basic variables), and A^i denotes the vector of constraint matrix (column) coefficients associated with the variable x_i.[5] For the knapsack problem, we can treat the knapsack constraint in the preceding formulation as the single constraint of the model, and set each nonbasic variable to either its lower limit of zero or its upper limit of b_i. A basic solution will therefore have a single basic variable that is free to take any value between its lower and upper bound values. Denoting the basic variable index as β, then we have $c_B = c_\beta/b_\beta$, $\mathrm{B} = 1$, and $\mathrm{B}^{-1} = 1$. The reduced cost of a variable i therefore becomes equal to $c_i/b_i - c_\beta/b_\beta$ (note that the dual multiplier value, $c_B \mathrm{B}^{-1}$, takes the value c_β/b_β). The optimality conditions for the linear program consist of primal feasibility, dual feasibility, and complementary slackness. For the greedy algorithm, primal feasibility is ensured by construction, i.e., we stop inserting items when we reach the capacity of the knapsack, and all variable values are maintained between their bounds. Dual feasibility requires a nonpositive reduced cost for each nonbasic variable at its lower bound, and a nonnegative reduced cost for each nonbasic variable at its

---

[4]Suppose we have an $m \times n$ matrix A and an m-dimensional column vector d, both consisting of real numbers, and suppose $n \geq m$. Consider the system of equations $Ax = d$, where x is an n-dimensional vector of variables. Suppose that we fix the values of $n - m$ of these variables at zero, and that the columns of the A matrix associated with the remaining m variables are linearly independent. Then the unique solution of the remaining m equations with m variables is called a *basic solution*. If, in addition, all of the variables in this unique solution are nonnegative, then we call this a *basic feasible solution*. We can generalize the notion of a basic feasible solution to allow for simple upper bounds on variables, e.g., for cases in which we require $0 \leq x \leq u$, where u denotes an n-vector of simple upper bounds on the variable values. If this is the case, and if we fix each of $n - m$ variables at either zero or their upper bound values, then the unique solution of the remaining m equations with m variables (assuming the remaining m columns are linearly independent) is a basic feasible solution if all variable values are nonnegative and less than or equal to their corresponding upper bounds.

[5]Consider a linear program of the form $\max cx : Ax = d, 0 \leq x \leq u$, where A is an $m \times n$ matrix, x is an n-dimensional vector of decision variables, d is an m-dimensional column vector, and u is an n-dimensional vector of upper bounds on the decision variable values. For this linear program, an optimal solution exists at an extreme point of the feasible region defined by the constraint set $Ax = d, 0 \leq x \leq u$, assuming the problem is not infeasible or unbounded. Moreover, a one-to-one correspondence exists between extreme points and basic feasible solutions. If the matrix A has full row rank, then a basic feasible solution may be written as $Bx_B + Nx_N = d$, where B is an $m \times m$ non-singular basis matrix, $0 \leq x_B \leq u$ denotes the set of basic variables, x_N denotes the set of nonbasic variables, each of which takes a value equal to zero or its upper bound, and N denotes the set of columns of the matrix A associated with the nonbasic variables. Letting A^i denote the column of A associated with the decision variable x_i, the reduced cost of this variable is written as $c_i - c_B B^{-1} A^i$; it is easy to see that the reduced cost associated with a basic variable equals zero. In addition, it is straightforward to show that if the reduced cost is nonpositive for each nonbasic variable equal to zero and nonnegative for each nonbasic variable at its upper bound, then the corresponding basic feasible solution is an optimal solution for the linear program (see [14]).

upper bound. This is also ensured by the greedy algorithm, as $c_i/b_i \geq c_\beta/b_\beta$ for each variable i at its upper bound, and $c_i/b_i \leq c_\beta/b_\beta$ for each variable i at its lower bound, based on our method of sorting items in nonincreasing order of the ratio c_i/b_i and inserting them into the knapsack in this order, until the knapsack capacity is exhausted or we run out of items. The complementary slackness condition can be written as $(c_\beta/b_\beta)(\sum_{i \in I} y_i - K) = 0$. Clearly if we have exhausted the knapsack's capacity, then this term equals zero; if, on the other hand, we run out of items before the capacity is exhausted, then the slack variable becomes the single basic variable, and the condition is again satisfied because this variable's objective function coefficient equals zero.

Observe that Algorithm 2.1 will always produce an optimal solution with at most one variable that takes a value strictly between its lower and upper bounds (we will refer to such a variable as the fractional variable; note that this fractional variable is the single basic variable in an optimal basic feasible solution). A stronger result actually holds, as it is possible to show that every extreme point of the feasible region of the linear program satisfies this condition. To see this, note that the polyhedron associated with the preceding linear program LP2 is fully n-dimensional, and an extreme point, by definition, occurs at the intersection of any n (or more) linearly independent hyperplanes. Constraints (2.10) form a set of $2n$ linearly independent hyperplanes, and clearly any intersection of n of these forms an extreme point solution with no fractional variable values, if the corresponding solution also satisfies the knapsack capacity constraint. Consider, therefore, the set of extreme points such that the knapsack constraint (2.9) is satisfied at equality. In order to form an extreme point, this constraint must intersect with $n-1$ of the bounding constraints (2.10). A total of $n-1$ variables must therefore take values at one of their bounds for any such extreme point solution, which implies that at most one variable can take a value strictly between its lower and upper bounds (note that we may in some cases have degenerate extreme points at which the knapsack constraint (2.9) is tight and an additional n of constraints (2.10) are tight, although this does not violate the condition that at most one variable takes a value between its bounds).

The LP relaxation solution obtained via Algorithm 2.1 provides an upper bound on the optimal solution value of the 0-1 knapsack problem. A slight modification of Algorithm 2.1 provides a simple heuristic solution method for the 0-1 knapsack problem. We simply modify Step 2 of the algorithm to set $y_l = 0$ if the condition is met. That is, the item that would otherwise take a value between its lower and upper bounds now equals zero, and the resulting solution is feasible for the 0-1 knapsack problem. The following section shows that under certain mild conditions, this heuristic method is *asymptotically optimal* in the number of items under the condition that the knapsack capacity K increases with the number of items, n. We can potentially improve upon the performance of this heuristic by continuing to go down the list of items and trying to fit additional items in the knapsack, as shown in Algorithm 2.2.

Algorithm 2.2 0-1 knapsack problem heuristic.

1: Sort items in nonincreasing order of c_i/b_i, set counter $l = 1$, $\hat{K} = K$, and $y_i = 0$ for $i = 1, 2, \ldots, n$.
2: If $b_l \leq \hat{K}$, then set $y_l = b_l$ and $\hat{K} = \hat{K} - b_l$. Otherwise continue.
3: Set $l = l + 1$. If $l = n + 1$, stop with heuristic solution. Otherwise return to Step 2.

2.4 Asymptotically Optimal Heuristic

This section shows that the modified greedy algorithm, where we essentially throw out the otherwise fractional variable from the LP relaxation, provides an asymptotically optimal solution for the 0-1 knapsack problem as the number of items increases to infinity, assuming the knapsack capacity grows with the number of items. The following equation provides the difference between the optimal solution value of LP2 (Z^*_{LP}) and the heuristic solution value (Z_H), where f is the index of the variable that takes a fractional value in the optimal LP relaxation solution, and the items are sorted in nonincreasing order of c_i/b_i:

$$Z^*_{LP} - Z_H = \frac{c_f}{b_f}\left(K - \sum_{i=1}^{f-1} b_i\right) \leq c_f. \tag{2.11}$$

We assume that all c_i and b_i parameter values come from bounded sets, i.e., $c_i \in [\underline{c}, \bar{c}]$ and $b_i \in [\underline{b}, \bar{b}]$ for some positive and finite $\underline{c}$, $\underline{b}$, $\bar{c}$, and $\bar{b}$, which implies from (2.11) that $Z^*_{LP} - Z_H \leq \bar{c}$. Because $Z^*_{IP} \leq Z^*_{LP}$ and (2.11) holds, we also have $Z^*_{IP} - Z_H \leq \bar{c}$, where Z^*_{IP} is the optimal 0-1 problem solution value. Therefore, regardless of the problem size, the difference between the LP relaxation solution value (an upper bound on the optimal 0-1 solution value) and the heuristic solution value is bounded by a fixed constant. Asymptotic optimality implies that the difference between the optimal solution value (of the 0-1 problem) and the heuristic solution value, taken as a percentage of the optimal solution value, approaches zero as the problem size approaches infinity. That is, we require

$$\lim_{n\to\infty} \frac{Z^*_{IP}(n) - Z_H(n)}{Z^*_{IP}(n)} = 0, \tag{2.12}$$

where $Z^*_{IP}(n)$ is the optimal 0-1 problem solution value for a problem containing n items and $Z_H(n)$ denotes the corresponding heuristic solution value. Because we have $Z^*_{IP}(n) - Z_H(n) \leq \bar{c}$ for any n, we need to show that $\lim_{n\to\infty} Z^*_{IP}(n) = \infty$ in order to show that (2.12) holds. To show this, note that all ratios c_i/b_i are bounded on some interval $[\underline{r}, \bar{r}]$, where $\underline{r} = \min_{i\in I}\{c_i/b_i\} > 0$ and $\bar{r} = \max_{i\in I}\{c_i/b_i\} < \infty$, and we assume that $\lim_{n\to\infty} K_n = \infty$, where K_n is the knapsack capacity for the n-item problem. That is, the knapsack capacity grows as a function of the number of items in the problem instance n, i.e., $K_n = f(n)$ where $f(n)$ is some function such that $\lim_{n\to\infty} f(n) = \infty$. Next, consider the following problem, the solution of which provides a lower bound on the optimal 0-1 problem solution value.

$$\text{Maximize} \quad \sum_{i\in I} \underline{r} y_i \tag{2.13}$$

$$\text{Subject to:} \quad \sum_{i\in I} y_i \leq K_n, \tag{2.14}$$

$$y_i \in \{0, b_i\}, \quad i \in I. \tag{2.15}$$

Note that the above formulation is equivalent to LP2 with the additional restriction that each y_i takes a value at one of its bounds, except that each coefficient c_i/b_i is replaced by the minimum ratio value $\underline{r}$. Let $\underline{Z}_H(n)$ denote the heuristic solution value of the above problem using our proposed heuristic, and define $\underline{Z}^*_{LP}(n)$ as the corresponding optimal LP relaxation solution value. Now note that $Z^*_{IP}(n) \geq \underline{Z}_H(n) \geq \underline{Z}_{LP}(n) - \underline{r} = \underline{r}(K_n - 1)$, where this last equality holds because an optimal LP relaxation solution exists such that the knapsack constraint is tight, as we have $\sum_{i\in I} b_i > K$ and $\underline{r} > 0$. Because $\underline{r} > 0$ and $\lim_{n\to\infty} K_n = \infty$ by assumption, we have $\lim_{n\to\infty} Z^*_{IP}(n) \geq \lim_{n\to\infty} \underline{r}(K_n - 1) = \infty$, which implies that (2.12) holds. Thus, while a large knapsack capacity may require greater computing time for an exact solution (as a result of the pseudopolynomial time algorithm), the implication of this asymptotic result is that for problems where the parameters c_i and b_i are drawn from bounded distributions, the simple heuristic algorithm that rounds down the single fractional variable from the LP relaxation solution tends to perform quite well.

Example 2.1 Consider a 10-item knapsack problem with the reward and capacity consumption values shown in the table below, and such that the knapsack has a capacity equal to 15.

Item, i	1	2	3	4	5	6	7	8	9	10
c_i	4	8	3/2	4	9	5	6	2	10	12
b_i	1	3	2	2	5	2	4	1	4	7
c_i/b_i	4	2.67	0.75	2	1.8	2.5	1.5	2	2.5	1.7

An optimal solution for the LP relaxation inserts items 1, 2, 6, 9, 4, and 8 fully into the knapsack (in this order), as well as 0.4 of item 5. The optimal LP relaxation objective function value equals 36.6. The asymptotically optimal heuristic solution method, which sets $x_5 = 0$ instead of 0.4, has an objective function value equal to 33. Observe, however, that we are able to improve this solution by inserting item 3 into the knapsack, which results in an objective function value of 34.5. If we solve the 0-1 problem to optimality, the optimal solution has an objective function value equal to 36, with $x_1 = x_2 = x_5 = x_6 = x_9 = 1$ and $x_3 = x_4 = x_7 = x_8 = x_{10} = 0$. □

2.5 Fast Approximation Algorithm

While asymptotic optimality may be a desirable feature for the performance of an algorithm, additional approaches can be used to provide performance guarantees that are arbitrarily close to optimality for any problem size. When an algorithm can guarantee a solution that is within $\epsilon \times 100\%$ of the optimal solution value in polynomial time (as a function of the problem size and $1/\epsilon$ for a fixed $\epsilon > 0$), such an approach is called a *fully polynomial time approximation scheme* (FPTAS). We next discuss an FPTAS for the knapsack problem. While we characterized our pseudopolynomial time solution approach as $\mathcal{O}(nK)$, a worst-case bound can actually be stated as $\mathcal{O}(n \min\{Z_{IP}^*, K\})$ by using a different solution approach, which we next describe. Define (C, A) as a pair of numbers denoting feasible objective function and capacity consumption values for the knapsack problem, i.e., there exists some $S \subset I$ such that $C = \sum_{i \in S} c_i$ and $A = \sum_{i \in S} b_i \leq K$. We can generate pairs of feasible (C, A) values iteratively as follows. We begin with a list containing only the pair $(0, 0)$ and we will increase the value of the index i at each step, beginning with $i = 1$. At each step, for every existing pair (C, A), if $A + b_i \leq K$, we consider adding the pair $(C + c_i, A + b_i)$ to the list (if $A + b_i > K$, then the candidate pair is eliminated). We do not add a candidate pair (C', A') to the list if some other (C, A) already exists on the list such that $C \geq C'$ and $A \leq A'$. Similarly, if we add a new pair (C, A) to the list, we can eliminate any pair already on the list that is dominated by (C, A). We maintain the list in increasing order of A so that for any item (C, A) on the list at the end of Step i, C is the greatest objective function value possible for a capacity consumption of at most A (accounting for items $1, \ldots, i$) and A is the minimum capacity consumption required for an objective function value of C. By sorting the list in increasing order of A, at any given Step i we can terminate the evaluation of the list and move to the next value of i whenever $A + b_i > K$. After Step n we arrive at (C^*, A^*) with $C^* = Z_{IP}^*$ providing the optimal objective function value and A^* providing the minimum capacity consumption required to achieve this objective function value. The time required for an iteration is a linear function of the number of pairs on the list, which is no greater than $\min\{Z_{IP}^*, K\}$, which implies an overall complexity of $\mathcal{O}(n \min\{Z_{IP}^*, K\})$.

This result does not appear to be useful *a priori*, as we do not know Z_{IP}^* in advance. Note, however, that if we know that Z_{IP}^* is small (smaller than K in particular), this approach gives an improvement over our prior algorithm based on dynamic programming.[6] We might therefore consider scaling the objective function by some value D to reduce the optimal objective function value. Our new algorithm, however, relies on the integrality of the optimal

[6] In addition, if we could characterize Z_{IP}^* as a polynomial function of n, then we could use the resulting $\mathcal{O}(nZ_{IP}^*)$ as our worst-case performance bound, which in such a case would be polynomial in n.

objective function value, and dividing by an arbitrary value of D can destroy this property. Alternatively, we can solve an approximate version of the problem by dividing each objective function coefficient by D and rounding down to ensure an optimal integer objective function value, i.e., we replace each c_i by

$$\tilde{c}_i = \left\lfloor \frac{c_i}{D} \right\rfloor, \tag{2.16}$$

where $\lfloor x \rfloor$ is the greatest integer less than or equal to x, to obtain an approximation of the problem we want to solve. If we apply our new algorithm to this approximation of the problem, a solution can then be obtained in $\mathcal{O}(n \min\{Z_{IP}^*/D, K\})$ time. Let S_A denote the set of items such that $x_i = 1$ in the solution to the approximate problem, and let Z_A denote the original problem's objective function value when using the solution to the approximate problem, i.e., $Z_A = \sum_{i \in S_A} c_i$. Our goal then is to choose a suitable value of D that depends on a preselected tolerance ϵ such that

$$\frac{Z_{IP}^* - Z_A}{Z_{IP}^*} \leq \epsilon, \tag{2.17}$$

and such that the algorithm runs in polynomial time as a function of n and $1/\epsilon$. Because our $\tilde{c}_i$ values were obtained by rounding down, we have $Z_A \geq D \sum_{i \in S_A} \tilde{c}_i$, which implies that

$$Z_{IP}^* - D \sum_{i \in S_A} \tilde{c}_i \geq Z_{IP}^* - Z_A. \tag{2.18}$$

Letting S^* denote the set of items such that $x_i = 1$ in the optimal solution to the original 0-1 knapsack problem, we also have $\sum_{i \in S_A} \tilde{c}_i \geq \sum_{i \in S^*} \tilde{c}_i$ (because the set of items in S^* provides a feasible solution for the approximate problem, while the set of items in S_A maximizes the objective of the approximate problem), which implies

$$Z_{IP}^* - D \sum_{i \in S^*} \tilde{c}_i \geq Z_{IP}^* - D \sum_{i \in S_A} \tilde{c}_i. \tag{2.19}$$

Because $c_i < D(\tilde{c}_i + 1)$ as a result of rounding down, this implies

$$Z_{IP}^* - D \sum_{i \in S^*} \tilde{c}_i = \sum_{i \in S^*} c_i - D \sum_{i \in S^*} \tilde{c}_i < D|S^*|, \tag{2.20}$$

where $|S^*|$ denotes the cardinality of the set S^*. Combining (2.18), (2.19), and (2.20), we have

$$D|S^*| > Z_{IP}^* - Z_A. \tag{2.21}$$

Given a desired value of ϵ, if we then choose D such that $D|S^*| \leq \epsilon Z_{IP}^*$, we can ensure that our desired performance bound (2.17) holds. We therefore need to choose some $D \leq \epsilon Z_{IP}^*/|S^*|$. Because $Z_{IP}^* \geq c_{\max}$ (where $c_{\max} = \max_{i \in I}\{c_i\}$) and $|S^*| < n$, we have $Z_{IP}^*/|S^*| \leq c_{\max}/n$, and we can therefore set $D = \epsilon c_{\max}/n$ and achieve the desired result. We now have that $Z_{IP}^*/D = nZ_{IP}^*/\epsilon c_{\max}$, and because $Z_{IP}^* \leq nc_{\max}$ we also have $Z_{IP}^*/D \leq n^2/\epsilon$. Returning back to our complexity bound for solving the approximate problem, we now have an $\mathcal{O}(n^3/\epsilon)$ algorithm that guarantees a solution within $\epsilon \times 100\%$ of the optimal solution value of the original knapsack problem.

We can further improve on this bound by strengthening the lower bound of $c_{\max}$ for Z_{IP}^* as follows. Recall that in our solution of the LP relaxation for the knapsack problem, we sorted items in nonincreasing order of the ratio c_i/b_i and filled the knapsack to capacity in this order, with item f denoting the single fractional item. The heuristic solution we previously discussed simply rounds down this fractional variable to zero, providing a heuristic solution value (and lower bound) of $Z_H = \sum_{i=1}^{f-1} c_i$. Defining a new lower bound of $Z_L = \max\{Z_H, c_{\max}\}$, instead of setting $D = \epsilon c_{\max}/n$, we can set $D = \epsilon Z_L/n$. Clearly the LP relaxation solution provides an upper bound on Z_{IP}^* and this upper bound is less than or equal to $\sum_{i=1}^{f} c_i = Z_H + c_f$. By the definition of Z_L we have that $Z_H + c_f \leq Z_H + c_{\max} \leq 2Z_L$, which implies $Z_{IP}^* \leq 2Z_L$. With our new value of D, we now have that $Z_{IP}^*/D \leq 2n/\epsilon$, which provides an order of magnitude improvement in the worst-case performance bound for solving the approximate problem, resulting in $\mathcal{O}(n^2/\epsilon)$.

2.6 Valid Inequalities

Throughout this book, we will in certain cases consider the use of valid inequalities in helping to solve an MILP or 0-1 programming problem. While the use of valid inequalities in solving these problems is not a focal point, in some cases these valid inequalities are very intuitive and can help significantly in providing strong bounds on optimal solution values (or even an optimal solution) for difficult problems. Thus, the treatment of valid inequalities in this book is intended to provide a flavor for general techniques, terminology, and the usage of valid inequalities, and to help in understanding the unique structural properties of certain difficult problems.[7]

When we solve the LP relaxation of the knapsack problem, we are maximizing the objective function over a continuous feasible region that contains all feasible 0-1 points, i.e., over a region that is a superset of all feasible points for the 0-1 problem. If all x_i values at all extreme points of this feasible region take values of 0 or 1, then solving the LP relaxation also provides an optimal

[7] In this book we will not, therefore, cover issues such as lifting of valid inequalities, separation problems, and facet proofs. For more details on these and other issues related to polyhedral analysis, please see Nemhauser and Wolsey [87].

solution to the 0-1 problem. Unfortunately, however, this is typically not the case, as the extreme points of the LP relaxation feasible region often take fractional values. If we could shrink the feasible region of the LP relaxation without eliminating any feasible 0-1 points, and maximize the objective over this smaller region, then we would obtain at least as good an upper bound on the optimal 0-1 solution. In particular, if we could shrink this region to the smallest convex set that contains all feasible 0-1 points, and maximize the objective over the resulting polyhedron, then we could solve an LP and obtain the optimal 0-1 problem solution value. The smallest convex set that contains all feasible 0-1 points is called the *convex hull* of integer feasible solutions, and unfortunately, determining the set of inequalities that defines this convex hull is at least as hard as solving the 0-1 knapsack problem. For MILP problems in general, and 0-1 integer programming problems in particular, it is often possible to aid in this cause by introducing *valid inequalities* that by definition do not eliminate any feasible 0-1 points, but may be able to shrink the size of the LP relaxation feasible region. It is important then to have some measure of the quality or "strength" of a valid inequality in terms of its ability to help in defining the convex hull of integer feasible solutions.

The inequalities included in a minimal set of valid inequalities required to define a convex hull of integer feasible points correspond to *facets.* A *face* of the convex hull is a set of points of the convex hull that satisfy one or more of the valid inequalities that define the convex hull at equality. For example, given a cube in three space, an extreme point is a zero-dimensional face, the line segment formed where two sides of the cube meet is a one-dimensional face, and each side is a two-dimensional face. For an n-dimensional convex hull, a facet is an $(n-1)$-dimensional face. For the three dimensional cube example, the inequalities that define the sides of the cube form the (six) facets. When using valid inequalities to eliminate portions of an LP relaxation feasible region, if we can ensure that an inequality represents a facet, then we know that no other inequality exists that dominates this inequality. By "dominates" we mean that no other valid inequality exists such that all of the feasible 0-1 points that satisfy the facet-defining inequality at equality, plus some additional feasible 0-1 points, also satisfy this other inequality at equality. Thus, a valid inequality defining a facet cannot be moved closer to the convex hull or "rotated" without either eliminating at least one feasible 0-1 point or weakening the inequality.

In certain cases, even though a valid inequality is not a facet, it can serve to improve the computational time required for a branch-and-bound algorithm in two ways. First, it can reduce the upper bound provided by the LP relaxation, which can reduce the required size of the branch-and-bound tree. Second, it can reduce the number of alternative optimal solutions for a given LP relaxation by reducing the size of the LP feasible region. Identifying inequalities that reduce the size of the LP relaxation feasible region without eliminating optimal solutions can be done in a number of ways, and we will consider deriving them by studying a problem's special structure and using logic to specify conditions that feasible and/or optimal solutions must satisfy. For example, for the knapsack problem, consider two items k and l such that item k takes up no more capacity than item l and has at least as high an objective function coefficient value, i.e., $c_k \geq c_l$. Suppose we have an optimal solution with $x_l = 1$ and $x_k = 0$. Then it is feasible to replace item l with item k, and the resulting objective function is at least as high as the original value. Therefore, either alternative optimal solutions exist, or we have a contradiction to the optimality of the original solution. This implies that the inequality $x_k \geq x_l$ is valid for the knapsack problem, and does not eliminate an optimal solution.

The most commonly used inequalities for improving the solution time of knapsack problems are the so-called *knapsack cover inequalities.* These cover inequalities are also based on a very simple and intuitive idea. If we have some subset $S \subset I$ of items such that $\sum_{i \in S} b_i > K$, then no more than $|S| - 1$ of these items can exist in a feasible solution to the 0-1 problem. Such a subset S is called a knapsack cover, and the associated knapsack cover inequality is written as

$$\sum_{i \in S} x_i \leq |S| - 1. \tag{2.22}$$

A cover set S is a minimal cover if the removal of any one item from S will result in a set whose capacity consumption values sum to a value less than or equal to K. If we index items in nonincreasing order of capacity consumption b_i, we can extend the cover inequalities associated with a minimal cover. Given a minimal cover S, consider the set $E(S)$ consisting of the union of S and all items not in S with capacity consumption values greater than or equal to $\max_{i \in S}\{b_i\}$. Given that no more than $|S| - 1$ elements of S can fit in the knapsack, then clearly if we replace any p elements of S (with $0 < p \leq |S|$) with p elements from $E(S) \backslash S$, no more than $|S| - 1$ elements of the resulting set will fit in the knapsack, which implies that the following inequality, which is stronger than (2.22), is also a valid inequality for the knapsack problem:

$$\sum_{i \in E(S)} x_i \leq |S| - 1. \tag{2.23}$$

In certain instances, inequalities of the form (2.23) form facets of the convex hull of integer solutions for the knapsack problem. If, for example, $S = I$ (and therefore $E(S) = S$), then it can be shown that (2.23) represent facets. Additionally, if $E(S) = I$ and the removal of two of the largest items from S along with the addition of a largest overall item forms a set that is capacity feasible (i.e., if S contains the r elements indexed $\{i_1, i_2, \ldots, i_r\}$ and $S \backslash \{i_1, i_2\} \cup \{1\}$ forms a set of capacity feasible items), then the resulting inequality of the form (2.23) represents a facet. Demonstrating that a valid inequality represents a facet of an n-dimensional convex hull of integer feasible solutions can be achieved by showing that n affinely independent points of the convex hull satisfy the inequality at equality. The points $x^1, x^2, \ldots, x^k \in R^n$ are affinely independent if $\lambda_1 = \lambda_2 = \cdots = \lambda_k = 0$ is the unique solution to $\sum_{i=1}^k \lambda_i x^i = 0$ with $\sum_{i=1}^k \lambda_i = 0$. A set of $n+1$ affinely independent points in n-space can be viewed as a simplex, i.e., if we have a set of n linearly independent points in n-space, this set of points is also affinely independent, and the addition of a point to this set that is not an affine combination of these n points also forms an affinely independent set of points.[8] The reader can verify that by definition there can be no more than $n+1$ affinely independent points in n-space.

Example 2.2 Consider the problem in Example 2.1. The optimal LP relaxation solution has an objective function value of 36.6, with variable values $x_1 = x_2 = x_4 = x_6 = x_8 = x_9 = 1$, $x_5 = 0.4$, and $x_3 = x_7 = x_{10} = 0$. The set of items $S = \{2, 4, 5, 6, 9\}$ forms a minimal cover, with a maximum constraint coefficient value of $b_5 = 5$; thus, the inequality

$$x_2 + x_4 + x_5 + x_6 + x_9 + x_{10} \leq 4,$$

is valid for the convex hull of solutions (note that because $b_{10} \geq b_5$, item 10 is a member of $E(S)$). When this inequality is added to the LP relaxation, the optimal solution value equals 36.33, with variable values $x_1 = x_2 = x_6 = x_8 = x_9 = 1$, $x_4 = 0.33$, $x_5 = 0.67$, and $x_3 = x_7 = x_{10} = 0$. Next note that the set of items $S = \{1, 2, 5, 6, 8, 9\}$ forms a minimal cover with $E(S) = S \cup \{10\}$. Therefore, the inequality

$$x_1 + x_2 + x_5 + x_6 + x_8 + x_9 + x_{10} \leq 5,$$

is valid for the convex hull of solutions. When this inequality is added to the LP relaxation (in addition to the prior inequality), the optimal solution equals 36.0, with $x_1 = x_2 = x_6 = x_9 = 1$, $x_4 = x_8 = x_{10} = 0.5$, and all other variables equal to zero. We know from Example 2.1 that this is an alternative optimal solution of the knapsack problem instance over the convex hull of integer-feasible solutions. An additional minimal cover set that will drive the LP relaxation solution to all integer values is the set $S = \{2, 6, 9, 10\}$, with an associated cover inequality $x_2 + x_6 + x_9 + x_{10} \leq 3$. When this inequality is added to the LP relaxation (along with the two previous inequalities), the resulting solution value remains at 36.0, with $x_1 = x_2 = x_5 = x_6 = x_9 = 1$ and all other variables equal to zero. □

[8] An affine combination of the n vectors $x^1, x^2, \ldots, x^n$ in $\mathbb{R}^n$ is a vector equal to $\sum_{i=1}^n \lambda_i x^i$ for any combination of the scalars $\lambda_1, \lambda_2, \ldots, \lambda_n$ such that $\sum_{i=1}^n \lambda_i = 1$.

2.7 Review

The knapsack problem is a singly-constrained 0-1 integer programming problem with applications in numerous fields. In general we view this problem as the assignment of some subset of a set of non-divisible items to a single constrained resource, where each item has a value, and both the resource capacity and the capacity consumption of each item can be measured using a single dimension. We presented an exact pseudopolynomial time dynamic programming formulation, and studied the linear programming relaxation for use in a branch-and-bound algorithm, a common approach for mixed integer linear programming problems. We also considered a simple but asymptotically optimal heuristic solution method, as well as a fast approximation algorithm that meets a predetermined tolerance on optimality in polynomial time (as a function of the inverse of this tolerance level, and for a fixed tolerance level). Finally, we considered some simple valid inequalities that can be used to tighten the LP relaxation of the 0-1 knapsack problem, and used this opportunity to discuss some general elements of the use of valid inequalities in solving MILP problems. Dynamic programming methods for the knapsack problem began with Bellman [17]. The fast approximation algorithm in Section 2.5 is due to Lawler [72], and a discussion on the valid inequalities in Section 2.6 can be found in Nemhauser and Wolsey [87].

Exercises

Ex. 2.1 — For the problem $\min\{\sum_{i\in I} c_i x_i : \sum_{i\in I} b_i x_i \geq K, 0 \leq x_i \leq 1, i \in I\}$, assuming a feasible solution exists (i.e., $\sum_{i\in I} b_i \geq K$), show that we can assume without loss of generality that all problem parameters are positive.

Ex. 2.2 — For the problem in Exercise 2.1, derive the LP optimality conditions and state an algorithm that solves this problem optimally.

Ex. 2.3 — For the problem $\min\{\sum_{i\in I} c_i x_i : \sum_{i\in I} b_i x_i \geq K, x_i \in \{0,1\}, i \in I\}$, provide an asymptotically optimal solution approach and demonstrate its asymptotic optimality.

Ex. 2.4 — For the problem described in Exercise 2.3, suggest how to derive a valid inequality of the form $\sum_{i\in I} x_i \geq k$ for some positive integer k, and demonstrate the validity of the resulting inequality.

Ex. 2.5 — For the problem in Examples 2.1 and 2.2, determine if additional minimal cover sets exist, and, if so, identify the corresponding sets S and $E(S)$ and associated minimal cover inequalities.

Ex. 2.6 — Propose a method for identifying a violated cover inequality for the 0-1 knapsack problem, given a solution to the LP relaxation (identifying the most violated such inequality is called the *separation problem*). Is this separation problem generally difficult? (You may consult the literature in answering this question.)

3

Set Covering, Packing, and Partitioning Problems

3.1 Introduction

Chapter 2 considered the problem of allocating multiple items to a single resource, i.e., the knapsack. In the knapsack problem, we have a set of items, from which we need to select a single subset of these items to allocate to the knapsack, with a goal of maximizing the value of the subset of items in the knapsack. We can thus view the knapsack problem as one of *packing* the knapsack with a subset of items that provides the highest value. A more general version of a so-called packing problem would permit selecting multiple subsets of items such that the subsets are disjoint (i.e., no single item is contained in multiple selected subsets) and each selected subset is allocated to a different resource. A distinguishing feature of these packing problems lies in the fact that the union of the subsets selected is not required to contain every item in the original set. In contrast, if we retain the requirement that the selected subsets must be disjoint, but require in addition that each item in the original set must be contained in some selected subset, then instead of a packing problem, we have a *partitioning* problem. For partitioning problems, our goal may again be to maximize the total value of all selected subsets, or we may cast the problem in terms of the cost associated with allocating a subset to a resource, in which case the objective is to minimize the total cost associated with the chosen subsets. Taking this a step further, if we remove the requirement that the selected subsets must be disjoint (but retain the requirement that each of the original items must be included in at least one selected subset), then we have a so-called *covering* problem. In this version of the problem, the solution must be such that each item is allocated to (or covered by) at least one resource.

This chapter discusses this class of covering, packing, and partitioning problems, for which numerous applications exist in operations contexts. For example, the set of items may correspond to production jobs, package deliveries, or customers, while the resources may correspond to production lines or machines, delivery vehicles, or service facilities. While, in general, this problem class may consider non-identical resources, we will restrict our analysis in this chapter to situations in which the multiple available resources are identical, and therefore the value or cost associated with a subset of items does not depend on the resource to which the subset is allocated (in principle, the set covering, packing, and partitioning problems we discuss in this chapter do not require explicitly defined resources, and the value or cost of a subset may depend only on the collection of items in the subset; in operations settings, we often have in mind a set of identical resources, such as machines, vehicles, or facilities to which a subset of items will be allocated). The following chapter on the generalized assignment problem (GAP) considers a particular type of set partitioning problem in which an item's cost and/or resource capacity consumption may depend on the resource to which the item is assigned.

Observe that if the contribution of an item to a subset's value or cost is independent of the other items contained in the subset, and each resource has unlimited capacity, in terms of the number of items a resource may accommodate, then the problem is trivial. In the packing case, we create a single subset containing all items with positive value. In the partitioning and covering cases, we create a single subset containing all items (assuming that, for the covering problem each item has an associated cost, and the objective consists of minimizing cost). Thus, for non-trivial problem instances, we implicitly assume that restrictions must exist on the number of items that a resource may accommodate and/or that the contribution of an item to a subset's cost or value is not independent of the other elements contained in the subset.

3.2 Problem Definition and Formulation

We consider a set I of items, such that the cost (or value) associated with item $i \in I$ equals c_i, and the item consumes b_i units of capacity of some standard resource. Let us also assume that we have a set J containing an infinite number of available resources, where each such resource has a capacity equal to K. Then it is feasible to allocate a subset $S \subseteq I$ to a resource if and only if $\sum_{i \in S} b_i \leq K$, and we let $\mathcal{F}$ denote the set of all subsets of I that are feasible for the problem we intend to solve. Let C_S denote the total cost associated with the subset S, and suppose that it is possible to enumerate all feasible subsets $S \in \mathcal{F}$ and associated costs C_S. We can characterize the contents of a subset $S \in \mathcal{F}$ using a column vector $a_S \in \mathbb{R}^{|I|}$ such that the ith element of a_S equals one if item i is contained in subset S, and equals zero otherwise. We also define a binary variable y_S for each $S \in \mathcal{F}$, which takes a value of one if we select subset S, and takes a value of zero otherwise. Letting $e^{|I|}$ denote an $|I|$-dimensional column vector such that each element equals one, we can formulate the set partitioning problem (SP) as follows.

$$\textbf{[SP]} \qquad \text{Minimize} \qquad \sum_{S \in \mathcal{F}} C_S y_S \tag{3.1}$$

$$\text{Subject to:} \qquad \sum_{S \in \mathcal{F}} a_S y_S = e^{|I|}, \tag{3.2}$$

$$y_S \in \{0, 1\}, \qquad S \in \mathcal{F}. \tag{3.3}$$

The objective (3.1) minimizes the total cost of all selected subsets, while the constraint sets (3.2) and (3.3) ensure that each item $i \in I$ is included within a single subset. A change in the equality sign in (3.2) to a less-than-or-equal-to sign ($\leq$) produces the set packing formulation, while a change to a greater-than-or-equal-to sign ($\geq$) gives the set covering formulation. As a result, the set packing and covering problems are both relaxations of the set partitioning problem.

The set partitioning problem falls in the class of problems which are $\mathcal{NP}$-Hard in the strong sense, which implies that we cannot obtain a polynomial time *or* pseudopolynomial time algorithm for this problem unless $\mathcal{P} = \mathcal{NP}$. Despite this, researchers have found that the value of the LP relaxation of this problem is often very close to the problem's optimal solution value, i.e., the LP relaxation is typically very "tight." Because of this, a branch-and-bound algorithm for SP is often not impractical, and operations researchers have cast numerous problem classes as set partitioning problems. In fact, Bramel and Simchi-Levi [19] showed that the LP relaxation value asymptotically approaches the optimal solution value of SP with probability one when C_S corresponds to the minimum vehicle route length required to serve customers in the set S, if all customer locations come from a set of independent random variables with compact support (and when customer demands are independent and are drawn from a compact, continuous distribution).

3.3 Solution Methods

If $C_S = C$ for some scalar C and for all $S \in \mathcal{F}$, then the resulting SP formulation is equivalent to the well-known *bin packing problem*, which seeks the minimum number of bins of standard size (or capacity) K required to accommodate each item in I (this common usage of the term *packing* is unfortunately not consistent with our definition of the set packing problem). Bin packing problems arise in numerous operations settings, including production line balancing problems in which a set of tasks must be divided among workstations, each of which has a given amount of daily work capacity, and the goal is to minimize the number of such workstations. Single-dimensional cutting stock problems, in which standard-sized resources (such as pieces of fabric or steel pallets) are divided into smaller pieces to satisfy individual demands for a material, can also be cast as bin packing problems when the objective consists of minimizing the number of standard-sized resources required to meet a given set of demands. The following subsection will discuss some well-known heuristic solution methods for the bin packing problem. Following this we will consider the more general version of SP formulated in the previous section, where each subset S may have a unique value of C_S, and where this value may depend on the collection of elements in S.

3.3.1 Bin packing heuristics

This section discusses four well-known and simple (but generally powerful) heuristic methods for solving bin packing problems. Each of these methods starts with a single open bin (bin 1) and proceeds down a list of items that must be allocated to bins, only opening a new bin when none of the existing open bins can accommodate the next item on the list. We use the convention of indexing the nth bin opened as bin n. In the most basic version of these heuristic approaches, the list of items is processed in an arbitrary order, and an item is inserted in the lowest indexed bin in which it will fit, given all prior assignments of items to bins (if none of the previously opened bins can accommodate the item, a new bin is then opened). This greedy heuristic approach is referred to as the *first-fit* method.

Instead of inserting the next item in the list into the lowest indexed open bin, suppose we insert the item into the bin that will have the lowest remaining capacity after inserting the item, among all open bins into which the item can fit (again, if the item cannot fit in any of the open bins, a new one is opened). The resulting approach, called the *best-fit* method, requires more effort than the first-fit method on average, as the item's size must be compared with the remaining capacity of all open bins at each step. Given a set of items I, let $B^*(I)$ denote the minimum number of bins required to accommodate all items in I. Garey, Graham, Johnson, and Yao [53] showed that both the first-fit and best-fit methods produce heuristic solutions that are bounded by $\lceil 1.7B^*(I) \rceil$, while Simchi-Levi [107] provided a bound of $1.75B^*(I)$ (where $\lceil x \rceil$ denotes the *ceiling function*, which simply requires rounding a real-valued x up to the smallest integer greater than or equal to x).

Improvements on these bounds are made possible by processing items in the set I in nonincreasing order of item sizes, instead of processing them in an arbitrary order. When this approach is applied with the first-fit method, the resulting heuristic is known as the *first-fit decreasing* heuristic. Similarly, when applied to the best-fit method, the resulting heuristic is called the *best-fit decreasing* heuristic. Baker [9] showed that the performance of both the first-fit decreasing and best-fit decreasing methods on a set of items I is bounded by $(11/9)B^*(I) + 3$, while Simchi-Levi [107] later showed that it is bounded by $1.5B^*(I)$.

Example 3.1 Consider an assembly line balancing problem in which no precedence constraints exist on operations, and the maximum amount of daily work content at a workstation equals 8 hours. The table below shows the daily work content (in hours) for each of 24 operations. We wish to allocate operations to workstations in order to minimize the total number of workstations. The resulting problem is a bin packing problem. Note that a simple lower bound on the required number of workstations equals $\lceil \sum_{i \in I} b_i/8 \rceil = 8$.

Item, i	1	2	3	4	5	6	7	8	9	10	11	12
b_i	1	4	5	6	3	2	1/2	1/4	3/4	1	1/2	3
Item, i	13	14	15	16	17	18	19	20	21	22	23	24
b_i	2	4	2	4	5	3/2	7/2	5/2	4	1	6	3/4

The reader may verify that both the first-fit and best-fit heuristics (processing items in the order they appear in the table) lead to a solution with 9 workstations, while the first-fit decreasing and best-fit decreasing heuristics produce a solution with 8 workstations (which must be an optimal solution because of the above lower bound calculation). □

3.3.2 Column generation and the set partitioning problem

In this section we use the set partitioning problem to illustrate a powerful technique for solving certain classes of difficult combinatorial optimization problems. This method, known as the column-generation method, is often applied to set partitioning problems (as well as the closely-related set packing and set covering problems). We begin by recognizing the fact that if we are given a set of items I containing $|I|$ elements, then the set of all possible subsets of I, denoted as $\bar{S}$, grows exponentially in $|I|$, and the set of all possible feasible subsets $\mathcal{F} \subseteq \bar{S}$ may grow exponentially as well. Thus, the SP formulation provided in the previous section is quite impractical because we require a column and a variable y_S for each possible feasible subset, i.e., the number of columns and corresponding binary variables grows exponentially in $|I|$ as well. The same is therefore implied if we wish to solve the linear relaxation of SP, i.e., the problem in which (3.3) is replaced by the requirement that $y_S \in [0, 1]$ for all $S \in \mathcal{F}$.

Despite this apparently daunting fact, the properties of optimal solutions for linear programs allow for techniques by which we can solve the LP relaxation of SP in practice with relative efficiency. In particular, note that an optimal basic feasible solution for the LP relaxation of SP contains $|I|$ basic variables (see Footnote 4 on Page 13 for the definition of a basic feasible solution). Moreover, a feasible solution to SP cannot contain more than $|I|$ binary variables equal to one. Thus, we are guaranteed to have an optimal basic solution for the LP relaxation of SP that contains no more than $2|I|$ positive variables. In other words, the number of columns that we actually need to include in an SP formulation that produces an optimal LP relaxation solution is relatively small (although we cannot generally know which columns we need *a priori*). That is, we could in principle essentially omit a very large number of columns from the formulation without loss of optimality, if we could determine a way to bypass columns that are not needed in finding an optimal solution. Recalling that the LP relaxation of SP tends to be "tight," the goal is then to incorporate the solution of the LP relaxation within a branch-and-bound scheme, in the hope that the gap between the LP relaxation and the optimal solution value of the instance of SP can be closed reasonably quickly.

Column-generation techniques provide a way to implicitly enumerate and bypass a large number of unneeded columns. In order to begin applying column generation, we need to start with a *restricted* version of the LP relaxation that contains a sufficient number of columns to guarantee that a feasible solution to the LP relaxation exists. Suppose, therefore, that we begin with a restricted version of the LP relaxation of SP containing only the columns of an $|I|$-dimensional identity matrix. That is, we begin with $|I|$ columns such that the ith such column contains a one in the ith row, and all other elements of the column equal zero. This restricted version of the problem also contains exactly $|I|$ binary variables $y_1, y_2, \ldots, y_{|I|}$, such that setting each of these variables equal to one provides a feasible solution for the LP relaxation of SP.

Beginning with this restricted formulation, we would like to identify *attractive* columns that we have not yet included in the formulation. We do this by first solving the LP relaxation of the restricted formulation. From this optimal solution to the initial restricted version of the problem, we can extract $|I|$ dual variables corresponding to the constraint set (3.2). Let π denote the vector of these dual variables (unrestricted in sign), where π_i denotes the ith such variable. With respect to the full formulation of the LP relaxation of SP (containing all possible columns), the optimal solution to our initial restricted LP relaxation is not likely to be optimal. However, with respect to the LP relaxation of the full formulation of SP, the optimal solution of the initial restricted LP relaxation does provide a basic feasible solution (BFS). This BFS and the corresponding dual variable values allow us to compute the *reduced cost* associated with any column that has not been included in the restricted version of the LP relaxation (if, given a subset S, we can compute the cost associated with this subset, C_S, which we assume throughout). The reduced cost of a column a_S can be computed as $C_S - \pi a_S$. If $C_S - \pi a_S \geq 0$ for all possible columns, then the current BFS is an optimal solution for the LP relaxation of SP. On the other hand, if some column a_S exists such that $C_S - \pi a_S < 0$, then the current BFS is not optimal (see Bazaraa, Jarvis, and Sherali [14]).

We do not wish to explicitly compute the reduced cost of every possible column that has not been included in the restricted LP relaxation. Instead, we wish to identify an *attractive* column that has not been included. That is, we wish to identify a column with a negative value of reduced cost (or show that no such column exists). To do this, we can solve a separate optimization problem with an objective of minimizing reduced cost. In this separate optimization problem, we now think of the column a_S as a set of $|I|$ binary decision variables, where $a_{S,i} = 1$ if item i is included in subset S, and $a_{S,i} = 0$ otherwise, for $i = 1, \ldots, |I|$. As a result, the cost of this as yet undetermined subset S depends on the composition of the column a_S, i.e., which elements of a_S equal one, and we let $C_S(a_S)$ denote a function that gives the cost associated with this column (with a slight abuse of notation).

Depending on the particular problem we are solving, a hypothetical column a_S corresponding to an element of $\bar{S}$ (the set of all possible subsets of I) may or may not be feasible. Recall that we let $\mathcal{F}$ denote the set of all *feasible* subsets of I for the problem we intend to solve. Then our separate optimization problem for determining the column with the most attractive reduced cost, which we refer to as the column *pricing problem*, PP, can be formulated as follows.

$$\textbf{[PP]} \qquad \text{Minimize} \qquad C_S(a_S) - \pi a_S \tag{3.4}$$

$$\text{Subject to:} \qquad a_{S,i} \in \{0, 1\}, \qquad i = 1, \ldots, |I|, \tag{3.5}$$

$$S \in \mathcal{F}. \tag{3.6}$$

If the optimal solution value of PP is nonnegative, then no column exists in the set of feasible columns $\mathcal{F}$ that may be added to the restricted version and can lead to an improved BFS, which implies that the current BFS is optimal for the LP relaxation of SP. If, on the other hand, the optimal solution value of PP is negative, this implies that we have identified a column that, if added to the restricted version of the problem, may lead to an improved BFS. This column is thus added to the restricted version of the problem and the corresponding new variable associated with the column is pivoted into the basis (and the simplex method is continued to determine a new BFS, with a corresponding new value of the vector of dual variables π). If this is the case, we then again solve PP using the new vector π, and repeat this process until we reach a point at which the optimal solution to PP is nonnegative, at which point the current BFS for the restricted version of the LP relaxation is optimal for the LP relaxation of SP. In this iterative process, the restricted version of the LP relaxation is often referred to as the *Master Problem*, while problem PP is known as the *Subproblem*.

Observe that if no column exists with a negative reduced cost, then problem PP must be solved to optimality in order to show that this is the case. However, when searching for an attractive column, it is not always necessary to solve problem PP to optimality. That is, for a given vector π at an iteration of the column-generation approach, many columns may exist with associated negative reduced costs. Proceeding back to the Master Problem with the addition of a new column only requires a column that "prices out," i.e., a column with a negative objective function value in problem PP. As a result, it is often possible to quickly solve problem PP heuristically (not to optimality), particularly in the early stages of the column-generation procedure, and obtain a column that prices out. If such a heuristic approach is applied and the resulting solution prices out, we can proceed by adding the associated column to the

Master Problem. If the heuristic approach fails to find a column that prices out, then problem PP must be solved to optimality to determine whether at least one additional column exists that prices out.

The difficulty of problem PP depends on both the definition of the set of feasible solutions $\mathcal{F}$ and the effort required in computing $C_S(a_S)$ for a set $S \in \mathcal{F}$. For example, a set partitioning approach may be used to solve a vehicle routing problem (VRP) instance (for more details, please see Section 9.6.1). In the vehicle routing problem, a subset S corresponds to an assignment of a subset of customers to a vehicle for deliveries departing from a distribution center (where $a_{S,i} = 1$ if customer $i \in I$ is assigned to the vehicle and $a_{S,i} = 0$ otherwise, for each $i \in I$). If K denotes the capacity of a standard truck, and customer i consumes b_i units of this capacity, then a subset S is feasible if and only if $\sum_{i \in S} b_i \leq K$, and the set of feasible solutions for PP may be represented by the constraint $\sum_{i \in S} b_i a_{S,i} \leq K$ (in addition to the binary requirement on the a_S vector). In addition, $C_S(a_S)$ corresponds to the minimum cost Hamiltonian Circuit[1] that includes all customers in S, as well as the distribution center. The resulting instance of PP is itself an $\mathcal{NP}$-Hard optimization problem (in the strong sense) in this case (the resulting problem corresponds to a so-called *prize-collecting traveling salesman problem*, wherein π_i serves as a prize for visiting customer i, but not all customers must be visited in the traveling salesman tour; please see Chapter 9 for a discussion of the traveling salesman problem, or the TSP).

As another example, suppose that S corresponds to a set of jobs, each of which must be assigned to one member of a set of identical machines, each with capacity K, and where b_i corresponds to the capacity consumption of job $i \in I$. As in the case of the VRP, a subset S is feasible if and only if $\sum_{i \in S} b_i \leq K$, and the set of feasible solutions for PP again corresponds to the constraint $\sum_{i \in S} b_i a_{S,i} \leq K$ (in addition to the binary requirement on the a_S vector). Suppose, additionally, that a fixed cost C is incurred for using a machine, and that the variable cost associated with a job is independent of the machine to which it is assigned (it is straightforward to generalize this approach to the case of non-identical machines, which we consider in more detail in the following chapter on the generalized assignment problem). Because the variable costs are machine-independent, and because we must assign each job to exactly one machine, the total variable costs for processing jobs are constant (and thus are the same for any feasible solution to the corresponding SP problem) and can therefore be ignored. In this case, the reduced cost associated with a subset S may be computed as $C - \pi a_S$, and the corresponding version of PP can be

[1] A Hamiltonian Circuit on a set of nodes in a network corresponds to a contiguous walk on the graph that enters and exits each of the nodes exactly once. Please see Chapter 9 for a discussion of the traveling salesman problem (TSP), which seeks a minimum cost Hamiltonian Circuit on all nodes in a complete graph, such that each arc has a given weight or cost.

formulated as the following knapsack problem.

$$\textbf{[PP(KP)]} \qquad \text{Maximize} \qquad \pi a_S \tag{3.7}$$

$$\text{Subject to:} \quad \sum_{i \in S} b_i a_{S,i} \leq K, \tag{3.8}$$

$$a_{S,i} \in \{0, 1\}, \qquad i = 1, \ldots, |I|. \tag{3.9}$$

We end this section by noting that the scheme we have described so far only provides a mechanism for solving the LP relaxation of SP. If this LP relaxation solution is indeed binary, then this solution also solves SP. In general, however, this does not yet give us the ability to solve SP. If the resulting LP relaxation solution is not binary, we cannot simply impose the binary requirements on the variables corresponding to the existing set of columns and claim that the resulting solution is optimal for the original SP problem. In other words, solving SP may require columns that were not generated in solving the LP relaxation of SP. The following section therefore describes a commonly used approach for using our ability to solve the LP relaxation approach within a branch-and-bound scheme that can lead to an optimal solution for an instance of SP.

Example 3.2 Consider the assembly line balancing (bin packing) problem from Example 3.1, and suppose we begin with the set of columns corresponding to the solution provided by the first-fit decreasing and best-fit decreasing heuristics (both solutions are the same). If we solve the corresponding LP relaxation of the SP formulation beginning with these columns, and continue to generate columns until no column has a negative reduced cost, the LP relaxation solution is obtained with an optimal solution value of 7.91 (Exercise 3.3 asks the reader to write out the pricing problem for the bin packing problem). Because the LP relaxation solution provides a lower bound on the optimal solution of SP, and because the optimal solution value must be an integer, we know that the optimal solution for this instance of SP is no lower than 8. Because the solution provided by the first-fit decreasing and best-fit decreasing heuristics (which is an upper bound on the optimal solution value of this instance of SP) equals 8, we conclude that we have an optimal solution. Observe that the optimal LP relaxation solution value for this problem instance is precisely equal to the simple lower bound of $\sum_{i \in I} b_i/8 = 7.91$. □

3.3.3 Branch-and-price for the set partitioning problem

The previous section provided a method for solving the LP relaxation of the set partitioning problem when the number of columns may prohibit explicitly writing out the entire problem formulation, i.e., when the number of feasible columns is huge. As noted previously, one reason for formulating a combinatorial optimization problem as a set partitioning problem lies in the fact that the set partitioning problem often has a small gap between the optimal binary solution and the problem's LP relaxation solution value. Because the LP relaxation solution value provides a lower bound (for a minimization problem) on the problem's optimal integer solution value, small gaps between the optimal integer solution value and the LP relaxation value often allow solution via a branch-and-bound approach without explicitly considering a large percentage of the possible nodes in a complete branch-and-bound tree.

Another perhaps equally important reason for casting a combinatorial optimization problem as a set partitioning problem, and using a column-generation approach, is the ability to efficiently solve the column pricing problem because of its special structure. For example, for the job-to-machine assignment example discussed in the previous section, the resulting pricing problem corresponds to a 0-1 knapsack problem. And although the 0-1 knapsack problem is an $\mathcal{NP}$-Hard optimization problem, as we discussed in the previous chapter, it does admit a pseudopolynomial time solution approach, which means that a fairly large class of instances can be solved with high efficiency. When a particular problem leads to a pricing problem with special structure, our goal is to exploit this special structure, and thus to solve the same type of pricing problem (or subproblem) at each iteration of the algorithm. For ease of exposition, the description of the branch-and-price approach that follows assumes a pricing problem of the form PP(KP).

The previous section described the "price" part of the branch-and-price approach, where column generation is used to solve the LP relaxation of a (typically restricted) version of the Master Problem at each node in a branch-and-bound tree. The "branch" part comes into play when the LP relaxation solution does not satisfy the binary requirements for the y_S variables. When this is the case, just as in a standard branch-and-bound algorithm, we create a node associated with each restricted problem instance, and partition the solution space such that the union of the spaces considered within each branch from a node provides a valid representation of the set of feasible solutions for the problem associated with the node, while also eliminating a part of the continuous feasible region containing the fractional solution associated with the node's Master Problem.

Suppose we solve the LP relaxation of SP and encounter an optimal solution containing fractional values for some subset of the y_S variables. In particular, suppose that $0 < y_S < 1$ for some $S \in \mathcal{F}$, and that we apply a branch-and-bound algorithm for solving SP. In a typical application of a branch-and-bound scheme, we would create two branches from the initial *root node* solution (corresponding to the LP relaxation of SP), one corresponding to the case in which $y_S = 1$ and one to the case in which $y_S = 0$. Clearly, because y_S was basic in the optimal solution of the LP relaxation of SP, we know that for the branch in which $y_S = 0$, the column corresponding to the set S prices out for the restricted problem associated with this branch. Thus, when we solve problem PP we are likely to obtain this same subset S as an optimal solution to problem PP. As a result, we obtain the same LP relaxation solution as we obtained at the *root node*, and we have made no progress with the algorithm. In order to make progress, we would need to keep track of the set S that corresponds to the variable on which we are branching and, if this solution is optimal for the pricing problem, we would need to consider the solution to PP with the second best value (as Vance, Barnhart, Johnson, and Nemhauser [114] note, at the nth level of the branch-and-bound tree, we potentially need to search for the nth best solution). Thus, the pricing problem we are solving needs to fundamentally change under this approach. Another way of viewing this phenomenon is by noting that when we branch, we effectively add a constraint to the problem SP (e.g., $y_S \leq 0$ for the branch corresponding to $y_S = 0$). This, in turn, alters the definition of the reduced cost with respect to the original LP relaxation problem.

As an alternative to branching on the y_S variables, researchers have developed successful strategies that effectively branch based on properties of the column vectors of SP instead. To illustrate this in the context of SP, we discuss the approach used by Vance et al. [114], originally proposed by Ryan and Foster [102]. Vance et al. [114] show that if we have a Master Problem with an optimal LP relaxation solution that is fractional (i.e., the values of two or more y_S variables are fractional), then an interesting property must hold. To describe this property, let $\tilde{S}$ denote the set of all columns included in the Master Problem formulation and, for any given pair of rows l and m, let $S_1(l,m) \subseteq \tilde{S}$ denote the set of columns (or subsets, indexed by S) such that both $a_{S,l} = 1$ and $a_{S,m} = 1$. Then, given an optimal fractional Master Problem solution, Vance et al. [114] showed that two rows l' and m' must exist such that $\sum_{S \in S_1(l',m')} y_S$ is strictly between 0 and 1 (otherwise, the Master Problem solution must be binary). If this is the case, then in order for (3.2) to hold, there must be two columns, say S' and S'', such that $a_{S',l'} = 1$ and $a_{S',m'} = 0$ while $a_{S'',l'} = 0$ and $a_{S'',m'} = 1$, with $y_{S'}$ and $y_{S''}$ both fractional. See Figure 3.1 for an illustration of the matrix elements associated with columns in the set $S_1(l',m')$ as well as the columns S' and S'' described. The resulting branching strategy focuses on the pair of items l' and m'. In particular, we partition the space of solutions so that on one branch we allow columns in $S_1(l',m')$ (as well as columns with a zero in both row l' and row m') and on the other branch we allow columns in which at most a single one exists in row l' or m'.

This is equivalent to branching on a pair of constraints. One branch requires $\sum_{S \in S_1(l',m')} y_S \leq 0$ while the second branch requires $\sum_{S \in S_1(l',m')} y_S \geq 1$. The first of these branches requires that items l' and m' cannot be in the same subset, while the second branch requires that items l' and m' are included in the same subset. In other words, the first branch only allows subsets S such that $a_{S,l'} = 0$ and $a_{S,m'} = 1$, $a_{S,l'} = 1$ and $a_{S,m'} = 0$, or $a_{S,l'} = 0$ and $a_{S,m'} = 0$. The second branch only allows subsets such that $a_{S,l'} = 1$ and $a_{S,m'} = 1$ or $a_{S,l'} = 0$ and $a_{S,m'} = 0$. After branching, any columns in the Master Problem that do not satisfy the branching constraints can be zeroed out. We must be careful, however, when solving the pricing problem PP to ensure that after branching, we do not generate any columns that violate the

	$S_1(l', m')$				S'		S''
Row l'	1	...	1	...	1	...	0
Row m'	1	...	1	...	0	...	1
	$0 < \sum_{S \in S_1(l',m')} y_S < 1$				$0 < y_{S'} < 1$		$0 < y_{S''} < 1$

FIGURE 3.1
In a fractional solution, rows l' and m' must exist such that $\sum_{S \in S_1(l',m')} y_S$ is fractional.

branching constraints. For the branch in which $\sum_{S \in S_1(l',m')} y_S \geq 1$, this is equivalent to combining items l' and m' into a single item that is either selected in the PP or is not. It is thus very easy to handle the associated PP in a manner consistent with the solution of a standard knapsack problem.

For the branch in which $\sum_{S \in S_1(l',m')} y_S \leq 0$, however, the resulting PP is considerably more difficult in general. One option would be to solve (at most) three instances of the PP. In the first instance, item l' is an available choice while item m' is not. In the second version, item m' is an available choice while l' is not. If the better solution between these two selects neither of these items, we are done. Otherwise we can solve a third instance of PP in which neither item l' nor item m' is an available choice. We cannot, however, afford to utilize this approach as we move down the branching tree, as accounting for all branching constraints associated with a node would lead to an algorithm with a number of steps that grows exponentially with the depth of the branch-and-bound tree. Vance et al. [114] discuss computational experience in which they use a depth-first strategy, under which they consider branches associated with the constraint $\sum_{S \in S_1(l',m')} y_S \geq 1$ wherever possible, and only consider branches associated with the constraint $\sum_{S \in S_1(l',m')} y_S \leq 0$ when necessary. They found that this strategy required solving relatively few of the latter problem types, which were solved using the mixed integer optimization solver MINTO.

Later, when we discuss the application of the branch-and-price approach to problems involving the assignment of items to a finite number of resources (e.g., the generalized assignment problem in Chapter 4 and the vehicle routing problem in Chapter 9), we will discuss more intuitive branching strategies that allow branching on the "underlying" assignment variables without fundamentally changing the structure of the associated pricing problem. The paper by Barnhart, Johnson, Nemhauser, Savelsbergh, and Vance [13] provides an excellent resource on the application of branch-and-price to large integer programming problems.

3.4 Review

Our analysis in this chapter focused primarily on the class of set *partitioning* problems, wherein each "item" must be contained in one of the selected subsets. As we will see in later chapters, the set partitioning model may be applied to several operations contexts, including the generalized assignment problem, lot sizing problems, facility location problems, and vehicle routing problems. The standard approach for each of these problems takes a given set of requirements, or demands, and minimizes the cost incurred in satisfying these requirements. In all of these problems, each requirement must be assigned to some form of a resource. In the lot sizing context, a period's demand (the item) must be assigned to one or more production setups; in the facility-location context, a customer's demand must be assigned to one or more facilities; in the vehicle routing context, each customer's delivery quantity must be assigned to one or more vehicles. When the solution structure is such that an optimal solution exists in which each requirement is allocated to exactly one resource (e.g., in the uncapacitated versions of the lot sizing and facility location problems or in the vehicle routing problem without delivery splitting), a set partitioning formulation may be used. The knapsack problem discussed in Chapter 2, in contrast, took a profit-maximizing approach, wherein the decision maker has the flexibility to choose which items to assign to the knapsack. In this chapter we saw how the knapsack problem may arise as a subproblem in the column-generation approach for solving the LP relaxation of the set partitioning problem, where, in the pricing subproblem, we seek the best selection of items (based on dual prices) to allocate to a resource.

The set *packing* problem, like the knapsack problem, does not require selecting (or "covering") every item, and its feasible region differs from that of set partitioning only in the inequality sign ($\leq$) used in the constraint set. As a result, set packing is a relaxation of the set partitioning problem, and we can use the branch-and-price approach described in this chapter (although the vector of dual multipliers, π, associated with constraint set (3.2) must be nonpositive). Clearly if each coefficient C_S of the objective (3.1) is positive, then the problem is trivial and no subset will be selected. Therefore, we can, without loss of generality, assume that each C_S value is negative, and, like the knapsack problem, we can take an approach that equivalently maximizes the collective value of all selected subsets (note that if we convert the objective (3.1) to a maximization problem of the sum of the negative C_S values, we then require a nonnegative vector of dual variables π, and the pricing problem will also change to a maximization objective). Thus, set packing problems are more appropriate for problems in which the decision maker wishes to maximize profit, and not all items must be selected.

In contrast, set *covering* problems consider the reverse direction ($\geq$) of the inequality in constraint set (3.2), and require that each item is "covered" by at least one resource. Such problems arise, for example, in emergency facility location problems, where each item corresponds to a customer, and a facility "covers" a customer if it is within some maximum allowable distance from the customer. The set covering problem is also a relaxation of the set partitioning problem, and again, the branch-and-price approach may be applied, although in this case, the associated vector of dual multipliers, π, must be nonnegative (the negative sign in front of π in the objective of the pricing problem PP therefore provides an incentive, or reward, for covering an item).

Exercises

Ex. 3.1 — Explain why we are guaranteed to have an optimal basic solution for the LP relaxation of SP that contains fewer than $2|I|$ positive variables.

Ex. 3.2 — Verify the results for the first-fit, best-fit, first-fit decreasing, and best-fit decreasing heuristics as applied to Example 3.1.

Ex. 3.3 — Write out the pricing problem for a set partitioning formulation of the bin packing problem.

Ex. 3.4 — Consider a feasible instance of SP (with a minimization objective), and let $Z_{=}^{SP}$ denote the optimal objective function value. Let $Z_{\geq}^{SP}$ and $Z_{\leq}^{SP}$ denote, respectively, the optimal objective function value associated with the corresponding instance of the set covering and set packing problems. What is the relationship, if any, between $Z_{=}^{SP}$, $Z_{\geq}^{SP}$, and $Z_{\leq}^{SP}$?

Ex. 3.5 — Write out a feasible instance of the set covering (packing) problem such that if we change the inequality $\geq$ ($\leq$) to equality $=$, the corresponding instance of the set partitioning problem has no feasible solution.

Ex. 3.6 — Using Figure 3.1, show that if $\sum_{S \in S_1(l',m')} y_S$ is fractional, then there must exist columns S' and S'' such that row l' contains a 1 in column S' and a zero in column S'' and row m' contains a 1 in column S'' and a zero in column S', with $y_{S'}$ and $y_{S''}$ both fractional.

4

The Generalized Assignment Problem

4.1 Introduction

The allocation of limited resources to various activities required in production and distribution of a good or service forms the basis of a majority of difficult operations planning problems. The knapsack problem, discussed in Chapter 2, considers the allocation of a single resource to multiple activities that essentially compete for the resource. In the knapsack problem class, we are free to omit items from the knapsack. In operations planning contexts where the knapsack capacity represents machine capacity and the items represent jobs, for example, this corresponds to leaving jobs unperformed. Many operations planning problems require that all members of a candidate set of jobs are assigned to some available resource, ensuring that all job processing requirements are met. In this chapter, we consider such a class of problems, where each member of a set of items must be assigned to some member of a set of available resources, and where each available resource has an associated capacity limit. The goal of this problem is to assign each item to a resource at minimum total cost while respecting resource capacities, where a cost is associated with the assignment of a given item to a specific resource. Unlike the knapsack problem, this generalized assignment problem (GAP) falls into the class of problems that are $\mathcal{NP}$-Hard *in the strong sense*, which implies that we are not likely to find even a pseudopolynomial time algorithm for its solution (unless $\mathcal{P} = \mathcal{NP}$). We will therefore focus on methods for determining good lower bounds on the optimal solution value, as well as high-quality heuristic solutions. These methods include Lagrangian relaxation, branch-and-price, and the use of an asymptotically optimal greedy heuristic approach.

As with the knapsack problem, the GAP considers a single dimension for measuring resource capacity. Relevant contexts for this problem class include, for example, the assignment of customer demands to production facilities, where facility capacity can be measured using a single dimension. The assignment cost assumption implies that the cost of assigning a customer to a facility is independent of all other customer-to-facility assignments. Thus, while we can formulate a problem requiring the assignment of freight deliveries to vehicles as a GAP, the resulting formulation is not able to explicitly account for costs that depend on the collective set of items assigned to a vehicle, e.g., the routing costs associated with the total distance traveled by each vehicle. (Chapter 9 discusses the use of a GAP formulation as part of a heuristic approach for solving vehicle routing problems.) On the other hand, applications in which multiple freight items share a common origin and destination pair, and must use multiple (possibly non-identical) vehicles or transportation modes, fit well within the GAP problem class. Numerous additional production applications lend themselves to the GAP model class, including single-dimension cutting stock and machine scheduling problems.

4.2 GAP Problem Definition and Formulation

Consider a set of items I, indexed by i, and a set of resources J, indexed by j. Resource j has an available capacity of K_j (we do not consider a time dimension in this problem class; thus, it can be viewed in some sense as a single-period problem for application to operations planning contexts). Assigning item $i \in I$ to resource $j \in J$ consumes b_{ij} units of resource j capacity and results in a cost of c_{ij}. Splitting the assignment of a single item among multiple resources is not permitted. We wish to assign each item in I to some resource in J without violating resource capacity limits, such that the total assignment cost is minimized. Our decision variables will therefore consist of binary assignment variables x_{ij}, where $x_{ij} = 1$ if we assign item i to resource j, and 0 otherwise. We can formulate the GAP as follows.

[GAP]

$$\text{Minimize} \quad \sum_{i \in I} \sum_{j \in J} c_{ij} x_{ij} \tag{4.1}$$

$$\text{Subject to:} \quad \sum_{i \in I} b_{ij} x_{ij} \leq K_j, \qquad j \in J, \tag{4.2}$$

$$\sum_{j \in J} x_{ij} = 1, \qquad i \in I, \tag{4.3}$$

$$x_{ij} \in \{0, 1\}, \qquad i \in I, j \in J. \tag{4.4}$$

The objective function (4.1) minimizes total assignment costs, while constraint set (4.2) ensures that we do not violate the capacity of any resource $j \in J$. Constraint set (4.3) requires assigning each item $i \in I$ to exactly one available resource. The name *generalized* assignment problem arises from the generalization of the special case in which $b_{ij} = 1$ for all $i \in I$ and $j \in J$, $K_j = 1$ for all $j \in J$, and $|I| = |J|$, which is the well-known standard assignment problem (or bipartite weighted matching problem), and which requires pairing items from the sets I and J at minimum cost (when $|I| = |J|$, the $\leq$ inequality in (4.2) is replaced with an equality sign). We note here also that if the objective is changed to a maximization and the equality in constraint set (4.3) is changed to less-than-or-equal-to ($\leq$), then the resulting formulation generalizes the multiple knapsack problem (in the multiple knapsack problem, the objective function and constraint coefficients are independent of the resource, i.e., $c_{ij} = c_i$ and $b_{ij} = b_i$ for each $i \in I$ and all $j \in J$; see Kellerer et al. [69]).

Example 4.1 Consider the following small instance of the GAP involving two resources and seven items. Resource 1 has a capacity of 5 units, while Resource 2 has 12 units of capacity available. The tables below provide the assignment costs, c_{ij}, and capacity consumption values, b_{ij}, for $i = 1, \ldots, 7$ and $j = 1, 2$.

c_{ij} values

Item, i	1	2	3	4	5	6	7
Resource 1	2	1	9	3	2	1	3
Resource 2	3	4	2	4	5	5	4

b_{ij} values

Item, i	1	2	3	4	5	6	7
Resource 1	3	3	2	4	2	3	2
Resource 2	6	6	6	1	2	1	1

The optimal LP relaxation solution when using the GAP formulation has an objective function value of 27.167, with the following variable values:

LP relaxation x_{ij} values

Item, i	1	2	3	4	5	6	7
Resource 1	1/3	1	1/2	0	0	0	0
Resource 2	2/3	0	1/2	1	1	1	1

The reader may verify that the optimal solution value for this instance of the GAP has an objective function value equal to 31, with $x_{21} = x_{31} = x_{12} = x_{42} = x_{52} = x_{62} = x_{72} = 1$, and all other variables equal to zero. The resulting integrality gap between the optimal LP relaxation solution and the optimal GAP solution equals $100 \times (31 - 27.167)/31 \approx 12.4\%$. We will return to this example after discussing additional techniques for solving the GAP. □

4.3 Lagrangian Relaxation Technique

Lagrangian relaxation is often applied to difficult optimization problems in order to obtain bounds on an optimal solution value. We apply Lagrangian relaxation by removing one or more constraints from the constraint set, and penalizing the violation of these constraints in the objective function. These relaxed constraints are typically complicating constraints, and their removal from the constraint set should lead to a relaxed problem that is solvable in acceptable time. For example, consider the following optimization problem, where the b_j and k values are scalars for $j = 1, \ldots, n$, D is an $m \times n$ matrix of real numbers, and d is an m-dimensional column vector.

$$\textbf{[P]} \qquad \text{Minimize} \quad \sum_{j=1}^{n} c_j x_j \tag{4.5}$$

$$\text{Subject to:} \quad \sum_{j=1}^{n} b_j x_j \geq k, \tag{4.6}$$

$$Dx \geq d, \tag{4.7}$$

$$x \in X. \tag{4.8}$$

The decision variables comprise the n-dimensional vector x, while the set X is defined as $X = \{x \in \mathbb{Z}_+^n : 0 \leq x \leq u\}$, where u is a finite n-vector of simple upper bounds on the decision variable values (and $\mathbb{Z}_+^n$ denotes the set of all nonnegative integer vectors in n-space).

Let Z_P^* denote the optimal solution value for problem P, and let Z_{LP}^* denote the optimal LP relaxation solution value, obtained by relaxing the integrality requirements imposed on the set X, i.e., by replacing (4.8) with $x \in Conv(X)$, where $Conv(\cdot)$ denotes the convex hull of integer feasible values in X. Note that $Z_{LP}^* \leq Z_P^*$ because the feasible region of the LP relaxation is a superset of that of the original problem P. Suppose that the first constraint (4.6) complicates the formulation, such that the removal of this constraint leads to an easier problem. Under the Lagrangian relaxation approach, we relax this constraint by removing it from the constraint set and penalizing its violation in the objective function. The amount of the penalty is denoted by λ, and this λ is called a dual or Lagrangian multiplier. Because constraint (4.6) is an inequality constraint, we can restrict λ to nonnegative values in order to penalize the objective function value when this constraint is violated. For nonnegative λ, the term $\lambda(k - \sum_{j=1}^{n} b_j x_j)$ penalizes the objective when (4.6) is violated (and "helps" the minimization objective when the constraint is satisfied), and the Lagrangian relaxation formulation becomes

$$\textbf{[LR(P)]} \qquad Z_{LR}(\lambda) = \text{Min} \ \sum_{j=1}^{n} c_j x_j + \lambda(k - \sum_{j=1}^{n} b_j x_j) \tag{4.9}$$

$$\text{Subject to: } Dx \geq d, \tag{4.10}$$

$$x \in X. \tag{4.11}$$

Given any feasible solution x' to problem P and a nonnegative λ, note that x' is feasible for LR(P), and the corresponding objective function value of LR(P) evaluated at x' is less than or equal to the objective function value of P evaluated at x', which implies $Z_{LR}(\lambda) \leq Z_P^*$ for any $\lambda \geq 0$. The solution of the Lagrangian relaxation problem therefore provides a lower bound on the optimal solution to problem P, and we would like for this lower bound to be as great as possible, which leads to the following Lagrangian dual problem,

LD.

$$\textbf{[LD]} \qquad Z_{LD} = \text{Max } Z_{LR}(\lambda) \tag{4.12}$$
$$\text{Subject to: } \quad \lambda \geq 0. \tag{4.13}$$

Let X' denote the convex hull of the feasible region defined by (4.10) and (4.11), i.e., $X' = Conv(X \cap \{x \in \mathbb{R}^n_+ : Dx \geq d\})$. Observe that for a fixed value of λ, the objective function of problem LR(P) is linear, and the optimal solution must therefore occur at an extreme point of the polyhedron defined by X' (although obtaining an explicit representation of X' may be a very difficult problem). Therefore, we can rewrite $LR(P)$ as

$$\textbf{[LR(P)]} \qquad Z_{LR}(\lambda) = \text{Min } \sum_{j=1}^{n} c_j x_j + \lambda(k - \sum_{j=1}^{n} b_j x_j) \tag{4.14}$$
$$\text{Subject to: } \quad x \in X'. \tag{4.15}$$

Next, consider the following alternative relaxation of problem P provided by Geoffrion [55].

$$\textbf{[R(P)]} \qquad Z_{R(P)} = \text{Min } \sum_{j=1}^{n} c_j x_j \tag{4.16}$$
$$\text{Subject to: } \sum_{j=1}^{n} b_j x_j \geq k, \tag{4.17}$$
$$x \in X'. \tag{4.18}$$

The relaxation R(P) is a linear program, and LR(P) is also a (Lagrangian) relaxation of R(P). This implies that $Z_{LR}(\lambda) \leq Z_{R(P)}$ for all $\lambda \geq 0$ which, in turn, implies that $Z_{R(P)}$ provides an upper bound on Z_{LD}. Let $\lambda^* \geq 0$ denote the optimal dual variable value associated with Constraint (4.17) for the linear program R(P). At an optimal solution x^* for problem R(P), we must have $\lambda^*(k - \sum_{j=1}^{n} b_j x_j^*) = 0$ by complementary slackness. Because x^* is feasible for LR(P), and because x^* and λ^* provide a solution to LR(P) with objective function value $Z_{LR}(\lambda^*) = Z_{R(P)}$, this implies $Z_{LR}(\lambda^*) = Z_{LD}$ (because $Z_{R(P)}$ is an upper bound on Z_{LD}). (Note that x^* solves LR(P) at $\lambda = \lambda^*$, which follows from the fact that LR(P) is equivalent to solving a Lagrangian relaxation of the linear program R(P), and the fact that a linear program has a Lagrangian duality gap of zero, i.e., the pair (x^*, λ^*) satisfies the so-called saddle point optimality conditions for LR(P); see Bazaraa, Sherali, and Shetty [15].)

The key result Geoffrion [55] showed was that solving the Lagrangian dual is equivalent to solving problem R(P). Because the feasible region of problem R(P), defined by constraints (4.17) and (4.18), is a subset of the LP relaxation feasible region for problem P, and because the objective function of problem R(P) is the same as that of problem P, we have $Z^*_{LP} \leq Z_{LD} \leq Z^*_P$, which tells us that the lower bound provided by solving the Lagrangian dual problem is at least as great as the one provided by the LP relaxation. Observe that if the feasible region defined by (4.17) and (4.18) is the same as the feasible region of the LP relaxation (which is equivalent to the condition that the linear relaxation of the feasible region of LR(P) has all integer extreme points[1]) then we have $Z_{LP}^* = Z_{LD} \leq Z_P^*$. This condition is known as the *integrality property*, and it implies that the lower bound provided by the Lagrangian relaxation is the same as that provided by the LP relaxation. When this condition does not hold, then $Z_{LP}^* < Z_{LD}$ is possible, and solving the Lagrangian dual problem may provide a better (higher) lower bound for problem P (a minimization problem).

Observe next that the formulation of LR(P) given by (4.14) and (4.15) implies that at any fixed λ, an optimal solution occurs at an extreme point of the polyhedron defined by X'. Because X' is a compact polyhedron defined by the intersection of a finite number of inequalities, the number of such extreme points is finite. For any extreme point of X', the objective of LR(P) ($Z_{LR}(\lambda)$) is a linear function of λ. Thus, we can view the function $Z_{LR}(\lambda)$ as the minimum among a finite number of linear functions of λ. It follows that $Z_{LR}(\lambda)$ is a piecewise-linear and *concave* function of λ, and Z_{LD} is obtained by maximizing a piecewise-linear concave function of λ on $\mathbb{R}_+$ (the nonnegative real line). The function $Z_{LR}(\lambda)$ is not, however, guaranteed to be differentiable at all values of λ. As a result, finding the value of λ that maximizes Z_{LD} is typically less than trivial.

Fisher [45] provides a now often-used method for solving the Lagrangian dual problem that employs the concept of subgradients for continuous functions that are not everywhere differentiable. A subgradient $\xi \in \mathbb{R}^n$ for a concave function $f(x)$ defined on a set $\mathcal{X} \subseteq \mathbb{R}^n$ at some point $\bar{x} \in \mathcal{X}$ satisfies the condition $f(x) \leq f(\bar{x}) + \xi(x - \bar{x})$ for all $x \in \mathcal{X}$ (technically, for a concave function, the value of ξ defined in this way should be referred to as a *supergradient*, although the term subgradient is often used in characterizing both concave and convex functions). Observe that if 0 is a subgradient of a concave function at $\bar{x}$, then by definition of the subgradient $f(x) \leq f(\bar{x})$ for all $x \in \mathcal{X}$, and $\bar{x}$ maximizes $f(x)$ over all $x \in \mathcal{X}$. The set of all subgradients at x is known as the subdifferential, denoted by $\partial f(x)$. If $f(x)$ is differentiable at $\bar{x}$, then $\partial f(x)$ consists of a singleton and is equal to the gradient of $f(x)$ at $\bar{x}$. Suppose the function $f(x)$ is piecewise linear and concave in x on a convex set $\mathcal{X} \subseteq \mathbb{R}$, with l breakpoints $\beta_1, \beta_2, \ldots, \beta_l$. Then for any value of the function $x \in \mathcal{X}$ that is not at a breakpoint, the subdifferential consists of the slope of the corresponding linear segment. For some breakpoint β_i, let $\partial^-(\beta_i)$ denote the slope of the function to the left of β_i, and let $\partial^+(\beta_i)$ denote the slope of the function to the right of β_i. Note that $\partial^-(\beta_i) > \partial^+(\beta_i)$ by

[1]To see why this is the case, observe that if the feasible region of LR(P) has all integer extreme points, then this implies $X' = Conv\,(x \in X : Dx \geq d) = Conv(x \in X) \cap \{x \in \mathbb{R}_+^n : Dx \geq d\}$. When this is the case, the intersection of X' with $\left\{x \in \mathbb{R}_+^n : \sum_{j=1}^n b_j x_j \geq k\right\}$, i.e., the feasible region of R(P), is identical to the feasible region of the LP relaxation of P.

the concavity of the function. Then the subdifferential at β_i consists of the interval $[\partial^+(\beta_i), \partial^-(\beta_i)]$. It is straightforward to show that if x is not an optimal solution for maximizing the concave function $f(x)$, then for a sufficiently small step in the direction of $\xi \in \partial f(x)$, the objective function value improves, which motivates the subgradient-search-based algorithm in Fisher [45] (for the minimization of a convex function, a step in the negative subgradient direction may lead to an improving solution).

Returning to LR(P), recall that Z_{LD} occurs at the maximum of a piecewise-linear and concave function of λ, where each line corresponds to an extreme point of the polyhedron defined by X'. For a given λ, this extreme point is obtained by solving $LR(P)$, and the associated linear function of λ has intercept $\sum_{j=1}^{n} c_j x_j^*(\lambda)$ and slope $k - \sum_{j=1}^{n} b_j x_j^*(\lambda)$, where $x^*(\lambda)$ solves LR(P) at the value of λ. As a result, $\xi(\lambda) = k - \sum_{j=1}^{n} b_j x_j^*(\lambda)$ corresponds to a subgradient of $Z_{LR}(\lambda)$ at λ and serves as a direction in which a step may be taken to improve the value of $Z_{LR}(\lambda)$ when 0 is not a subgradient at λ. Fisher [45] provides rules for a sequence of step sizes that guarantee theoretical convergence to the value of λ that maximizes $Z_{LR}(\lambda)$, leading to the value of Z_{LD}.[2]

Our brief discussion of Lagrangian relaxation used an example in which we relaxed a single constraint and the set of feasible solutions was discrete and finite. Generalizing the approach to the relaxation of multiple constraints and/or to problems with continuous variables and an infinite number of feasible solutions is straightforward, and our subsequent application of Lagrangian relaxation to the GAP will illustrate the case in which multiple constraints are relaxed. Later, in Chapters 5, 7, and 8, we will illustrate Lagrangian relaxation techniques for problems involving continuous variables (and an infinite number of feasible solutions) as well.

[2]As with methods that step in the gradient direction, subgradient optimization adjusts the value of λ by taking steps in the subgradient direction. Suppose that at iteration l, we use a Lagrangian multiplier equal to λ^l, and the corresponding Lagrangian relaxation solution equals $x^*(\lambda^l)$, with a subgradient value of $\xi(\lambda^l) = k - \sum_{j=1}^{n} b_j x_j^*(\lambda^l)$. Then we adjust the Lagrangian multiplier at iteration $l+1$ by setting $\lambda^{l+1} = \lambda^l + s^l \xi(\lambda^l)$, where s^l corresponds to some positive step size. Poljak [94] showed that as long as $\sum_{l=1}^{\infty} s^l \to \infty$ and $s^l \to 0$ as $l \to \infty$, the sequence of λ values will converge to λ^*, the value of λ that gives Z_{LD}. Held, Wolfe, and Crowder [66] suggested a rule for setting a step size that is commonly used in practice, where

$$s^l = \gamma_l \frac{\hat{Z} - Z_{LR}(\lambda^l)}{||\xi(\lambda^l)||^2}.$$

In this formula, $\hat{Z}$ is an upper bound on Z_{LD}, and γ_l is a scalar in the interval $(0, 2]$. Fisher suggests setting $\gamma_1 = 2$ and dividing this parameter by 2 if the value of $Z_{LR}(\lambda)$ has not improved during the past p iterations, where p is a parameter determined by the user. Because we cannot typically validate having found Z_{LD}, the procedure is usually terminated after some limit on the number of iterations. Note that if we relax multiple constraints, using a vector of Lagrangian multipliers λ, we apply the same step size to each element of the vector. If the step size leads to a negative value of λ for some element(s) of the vector (assuming we require $\lambda \geq 0$), then for each such element we set the corresponding element to zero, which gives the closest projection of the vector λ to the feasible set $\lambda \geq 0$.

4.3.1 Lagrangian relaxation for the GAP

In considering the application of Lagrangian relaxation to the GAP, we would like to obtain a relaxed problem that is solvable with relative efficiency, and that provides a stronger (higher) lower bound than the one obtained via the LP relaxation. We have two fundamental choices: relaxing the capacity constraints (4.2) or the assignment constraints (4.3) (we can choose to relax any subsets of these constraints, although we are looking for some resulting structure in the relaxed problem that we can exploit; because of this, it is useful to consider relaxing the entire set of either of these constraint sets). The first observation we can make is that if we relax the entire set of capacity constraints (4.2), the remaining set of constraints contains a matrix of coefficients that is *totally unimodular*, which implies that the so-called integrality property holds (see Exercise 4.4). That is, the assignment constraint set forms a feasible region such that all extreme points of the corresponding polyhedron are binary. As a result, the bound obtained via Lagrangian relaxation will equal that obtained by solving the original LP relaxation. Because of this, it will likely be more productive to relax the set of assignment constraints (4.3).

To this end, let λ_i denote a Lagrangian multiplier for each $i \in I$. The resulting Lagrangian relaxation formulation can be written as

[LR(GAP)]

$$\text{Minimize} \sum_{i \in I} \sum_{j \in J} (c_{ij} - \lambda_i) x_{ij} + \sum_{i \in I} \lambda_i \quad (4.19)$$

$$\text{Subject to:} \quad \sum_{i \in I} b_{ij} x_{ij} \leq K_j, \quad j \in J, \quad (4.20)$$

$$x_{ij} \in \{0, 1\}, \quad i \in I, j \in J. \quad (4.21)$$

For a given vector $\lambda \in \mathbb{R}^{|I|}$, the above problem decomposes into $|J|$ 0-1 knapsack problems (note that because the constraints we have relaxed are equality constraints, the λ_i variables are unrestricted in sign). Each of these knapsack problems can be solved in pseudopolynomial time using the dynamic programming approach discussed in Section 2.2. While a subgradient search algorithm may be applied to maximize the Lagrangian dual, Fisher, Jaikumar, and Van Wassenhove [48] suggest a multiplier adjustment method that provides strong bounds and converges finitely.

A primary concern in applying Lagrangian relaxation to the GAP lies in obtaining good feasible solutions as candidate upper bounds on the optimal solution of the GAP. A solution to LR(GAP) for any value of the vector λ may violate feasibility of the GAP itself in two ways. First, a given item $i \in I$ may be assigned to multiple resources in J. This type of infeasibility is easily resolved by removing the item from all resources except for the one with the lowest value of c_{ij}. Second, a given item $i \in I$ may have not been assigned to any resource at all. When this occurs, feasibility of the solution may be difficult to achieve. Ideally, each unassigned item would fit in the corresponding resource with the lowest value of c_{ij}. However, this cannot be guaranteed, and any heuristic approach to restoring feasibility by ensuring that each item is assigned to a single resource may fail. Ultimately the use of a branch-and-bound algorithm is required to ensure exact solution of the GAP. However, the stronger bounds obtained via Lagrangian relaxation at each node of the branch-and-bound tree can substantially reduce the solution time required (see Ross and Soland [97] and Fisher et al. [48] for details on such an approach).

Example 4.2 Returning to the problem in Example 4.1, if we maximize the Lagrangian dual, the optimal Lagrangian multiplier values are given by $\lambda_1 = 21$, $\lambda_2 = 22$, $\lambda_3 = 20$, $\lambda_4 = 4$, $\lambda_5 = 13$, $\lambda_6 = 5$, and $\lambda_7 = 14$. When using these Lagrangian multiplier values, the Lagrangian relaxation problem, LR(GAP), has an optimal objective function value equal to 31 (Exercise 4.3 asks the reader to verify this result). Note that this value provides a lower bound on the optimal solution of the GAP instance. In Example 4.1, we noted that the optimal solution of this instance of the GAP has an objective function value of 31. Thus, for this problem instance, Lagrangian relaxation enables closing the entire integrality gap of the LP relaxation lower bound from the GAP formulation. Although this is not guaranteed to always be the case, this example illustrates the potential for Lagrangian relaxation to provide strong lower bounds on the optimal solution value. □

4.4 Branch-and-Price for the GAP

Savelsbergh [104] developed a branch-and-price approach for solving the GAP, which he showed to be quite efficient for solving problems in which the number of items assigned to any resource is relatively small, but where the number of resources may be fairly sizeable. Thus, for problems in which $|I|/|J|$ is small (e.g., $|I|/|J| \leq 5$), reformulating the GAP as a restricted version of the set partitioning problem (SP) often permits efficient solution via branch-and-price. Formulating this restricted version of SP requires additional notation, which we next describe.

Instead of considering all feasible subsets $\mathcal{F}$ of the set of items I for a standard resource, as we did when formulating SP in Chapter 3, we are now concerned with the set of all feasible subsets that may be assigned to each of the resources in the set J. Therefore, we let $\mathcal{F}_j$ denote the set of all feasible subsets of items in I that may be assigned to resource $j \in J$. Let $S_j^k \in \mathcal{F}_j$ denote the kth such subset, and let a_j^k denote the associated column vector, where $a_{ji}^k = 1$ if item i is included in the kth feasible subset associated with resource j, and zero otherwise. A subset S_j^k is feasible for resource j if $\sum_{i \in S_j^k} b_{ij} \leq K_j$, and the cost associated with subset S_j^k equals $\sum_{i \in S_j^k} c_{ij}$, which we denote as C_j^k. Let y_j^k denote a binary decision variable equal to one if subset S_j^k is selected for resource j, and equal to zero otherwise, for each $j \in J$ and each subset $k = 1, \ldots, |\mathcal{F}_j|$.

Solving the GAP requires assigning each item $i \in I$ to a resource, as well as choosing at most one subset for each resource $j \in J$. Using the notation we have defined, we can formulate the GAP as a set partitioning problem with an additional constraint set as follows.

$$\textbf{[SP(GAP)]} \quad \text{Minimize} \sum_{j \in J} \sum_{k=1}^{|\mathcal{F}_j|} C_j^k y_j^k \tag{4.22}$$

$$\text{Subject to:} \sum_{j \in J} \sum_{k=1}^{|\mathcal{F}_j|} a_j^k y_j^k = e^{|I|}, \tag{4.23}$$

$$\sum_{k=1}^{|\mathcal{F}_j|} y_j^k \leq 1, \qquad j \in J, \tag{4.24}$$

$$y_j^k \in \{0,1\}, \qquad j \in J, k = 1, \ldots, |\mathcal{F}_j|. \tag{4.25}$$

The objective (4.22) minimizes the total cost of assignments to resources, while constraint set (4.23) ensures that every item is assigned to a single resource. Constraint set (4.24) permits selecting at most one subset for each resource $j \in J$. As with SP, it is likely to be impractical to enumerate all feasible subsets for each resource. Instead, we apply a branch-and-price approach for solving the SP(GAP). In order to do this, we need to begin with a subset of the feasible columns that ensures that a feasible solution exists for the problem's LP relaxation. One way to do this is to add an artificial dummy column consisting of all ones and with very high cost (corresponding to a dummy resource).

We next discuss how to solve the LP relaxation via column generation. Unlike the basic SP discussed in the previous chapter, the pricing problem formulation differs for each type of resource for the SP(GAP). As with the basic SP, we let π denote an $|I|$-dimensional vector of dual variables associated with constraint set (4.23). In addition, we require a nonnegative dual variable α_j for each constraint of the form (4.24). Then the reduced cost associated with a feasible subset S_j^k for resource j equals $C_j^k - \pi a_j^k - \alpha_j$, where $C_j^k = \sum_{i \in I} c_{ij} a_{ji}^k$. Letting c^j denote an $|I|$-dimensional vector of c_{ij} values, we can write $C_j^k = c^j a_j^k$. Then the pricing problem associated with resource $j \in J$, which we denote as PP_j, can be written as

$$\textbf{[PP}_j\textbf{]} \quad \text{Maximize} \quad \left(\pi - c^j\right) a_j \tag{4.26}$$

$$\text{Subject to:} \quad \sum_{i \in I} b_{ij} a_{ji} \leq K_j, \tag{4.27}$$

$$a_{ji} \in \{0,1\}, \qquad i = 1, \ldots, |I|, \tag{4.28}$$

where we have suppressed the subset index k, as it is unnecessary in the description of the pricing problem. As with the basic version of the SP discussed in the previous chapter, the pricing problem takes the form of a knapsack problem. Observe that given an optimal solution to PP_j, the reduced cost of the column generated equals the negative of the optimal objective function value minus α_j. In the worst case, when solving the LP relaxation of SP(GAP), we may have to solve a pricing problem for each resource $j \in J$. However, as before, we can begin by initially generating heuristic solutions quickly, which are likely to price out in the early iterations of the algorithm.

As noted in the previous chapter, if the LP relaxation of SP(GAP) is not binary at the root node, branching is required. As also discussed in the previous chapter, problems arise if we attempt a standard branching approach, e.g., for a fractional y_j^k, creating branches for the case of $y_j^k \leq 0$ and $y_j^k \geq 1$. For the GAP, the existence of a finite set of individual resources actually makes the description of a good branching strategy much easier than for the SP discussed in the previous chapter. Given a solution to the SP(GAP), we can map this solution back onto the *underlying* x_{ij} variables in the original GAP formulation; in particular, we have $x_{ij} = \sum_{k=1}^{|\mathcal{F}_j|} a_{ji}^k y_j^k$. Savelsbergh [104] showed that when a variable y_j^k is fractional, then some item i exists such that x_{ij} is fractional, i.e., there is an item i that is only partially assigned to resource j. The branching strategy chooses such an underlying variable for branching, instead of branching on the y_j^k variables in the SP(GAP) formulation.

In order to branch on an underlying variable, suppose we have identified an underlying variable that has been fractionally assigned to resource $j \in J$ in the optimal solution to the LP relaxation of the SP(GAP). We can then create a branch for the case of $x_{ij} = 0$ and one for the case of $x_{ij} = 1$. The branch for the case of $x_{ij} = 0$ corresponds to the formulation of the SP(GAP) when no columns are permitted such that $a_{ji}^k = 1$, i.e., a_{ji}^k must equal zero for all problems solved (including PP_j) when this branch is followed. In the pricing problem PP_j, this is accomplished by eliminating the variable a_{ji} from the formulation. The branch for the case of $x_{ij} = 1$ corresponds to the formulation of the SP(GAP) when all columns for resource j must have a one in the ith position, and all columns associated with other resources must have a zero in this position. In the pricing problem PP_j, the variable a_{ji} is fixed to one by subtracting b_{ij} from the right-hand side of (4.27), adding the constant c_{ij} to the objective, and solving the problem in the remaining variables. The pricing problems for all other resources must eliminate item i from consideration.

4.5 Greedy Algorithms and Asymptotic Optimality

Most so-called *greedy* algorithms proceed down a list of items, at each step making the (myopically) best available choice for the item being processed. For some problem classes, a particular greedy algorithm may even be optimal. For instance, for the continuous knapsack problem in Chapter 2, a greedy algorithm is optimal if we process items in nonincreasing order of the ratio of the objective function coefficient to the constraint coefficient for each item. When multiple constrained resources exist, as in the GAP, it is typically useful to consider a measure of not only the best choice for the next item on the list, but also some measure of the cost of not making this choice. In other words, when we process an item on the list, if the best choice for the item is no longer feasible because of previous choices made in the greedy algorithm, the second-best choice for the item may be available, and this would typically be the action taken under a greedy approach. If the difference between the best choice for the item and the second-best choice is large, then we are likely to, in a sense, pay a big penalty if the best choice is not available when the item is processed. It is, therefore, preferable to consider such items early in an algorithmic approach (i.e., these items should take higher priority). A greedy approach based on the value of this difference (or potential penalty) for each item is, therefore, not purely myopic at each step. Such an approach considers the implications of an inability to make the best choice for each item at every step in the algorithm.

To describe such a modified greedy approach for the GAP, we consider some measure, denoted as $f(i,j)$, of the cost of assigning item i to resource j. For example, if $f(i,j) = c_{ij}$, then this measure considers only the cost of assigning item i to resource j; alternatively, if $f(i,j) = c_{ij}/b_{ij}$, this measure considers the cost of the assignment relative to the amount of resource j capacity consumed by item i. Martello and Toth [83] suggested these measures for use in a greedy approach for the GAP, in addition to a few others. However, instead of processing items based on $f(i,j)$ values, Martello and Toth [83] suggested an approach that considers both the best and second-best choices for each item $i \in I$. To do this, they created a measure of the desirability of making the best assignment for each item $i \in I$, denoted by ρ_i. Letting j_i denote the first-best choice of resource for item i, i.e., $j_i = \arg\min_j f(i,j)$, then the desirability value for item i is computed as

$$\rho_i = \min_{l \neq j_i} f(i,l) - f(i,j_i), \tag{4.29}$$

i.e., the difference in cost between the best and second-best choice of resource. The modified greedy approach then processes items in nonincreasing order of desirability values. Romeijn and Romero Morales [95] provided a measure $f(i,j)$ that leads to powerful solution properties in the application and analysis of a modified greedy solution approach. This measure uses a vector $\alpha \in \mathbb{R}_+^{|J|}$ and defines the value of $f(i,j)$ as

$$f_\alpha(i,j) = c_{ij} + \alpha_j b_{ij}, \tag{4.30}$$

which, like the ratio c_{ij}/b_{ij}, permits accounting for the impact of an item's cost and resource consumption. However, $f_\alpha(i,j)$ has the added benefit of a resource-specific *penalty* for capacity consumption (via the vector of "resource capacity consumption penalties," α). Observe that (4.30) corresponds to the coefficient of the variable x_{ij} if we apply a Lagrangian relaxation scheme that relaxes the capacity constraints (4.2) of the GAP (note also the similarity between (4.30) and the reduced cost of the variable x_{ij} in the GAP formulation). The first important result demonstrated by Romeijn and Romero Morales [95] can be described as follows. Suppose that the LP relaxation of the GAP is feasible and non-degenerate, with an optimal dual multiplier vector corresponding to constraints (4.2) equal to $\alpha^D \leq 0$. If the modified greedy algorithm uses $\alpha = -\alpha^D$ in (4.30) and if $x_{ij} = 1$ in the LP relaxation solution of the corresponding instance of the GAP, then $x_{ij} = 1$ in the application of the modified greedy algorithm.

The reason this result is powerful is because it can be shown (see [18], [95]) that the number of items $i \in I$ that are split between resources in the LP relaxation solution is less than or equal to the number of tight resource capacity constraints, which is less than or equal to the number of resources $|J|$. To see this, observe that the GAP formulation contains the two constraint sets (4.2) and (4.3), in addition to the binary restrictions (4.4). The LP relaxation is obtained by replacing the binary restrictions with nonnegativity constraints (upper bounds on the x_{ij} variables are unnecessary in the LP relaxation, as they are implied by the nonnegativity and assignment constraints (4.3)). A basic solution[3] for the LP relaxation of the GAP therefore can contain at most $|I| + |J|$ positive variables (where $|I| + |J|$ is the number of capacity and assignment constraints). Define s (ns) as the number of jobs that are split (are not split) between resources, and let R_e denote the number of resources with excess capacity (and, therefore, positive slack variables) in a basic solution. Let τ denote the total number of positive variables associated with split jobs. If we define γ as the average number of resources to which a split item has been assigned, then $\gamma = \tau/s$, i.e., $\gamma s = \tau$. Then we can write the total number of positive variables in a solution as $\gamma s + ns + R_e$. Because $s + ns = |I|$, the number of positive variables may be written as $(\gamma - 1)s + |I| + R_e$, which is bounded by the maximum number of positive basic variables, $|I| + |J|$, i.e.,

$$(\gamma - 1)s + |I| + R_e \leq |I| + |J| \tag{4.31}$$

$$\Rightarrow (\gamma - 1)s \leq |J| - R_e. \tag{4.32}$$

Note that the right-hand side of (4.32), $|J| - R_e$, is the number of resources with zero slack in a basic solution. Letting R_t denote the number of resources with zero slack, we have $R_t = |J| - R_e$, and we can write (4.32) as

$$s \leq \frac{R_t}{\gamma - 1}. \tag{4.33}$$

Because we must have $\gamma \geq 2$ (because any split job must be assigned to two or more resources), this implies that the number of split items, s, satisfies $s \leq R_t \leq |J|$.

It is therefore possible to demonstrate the asymptotic optimality of the

[3] See Footnote 4 on Page 13 for the definition of a basic solution.

modified greedy heuristic when the number of resources is held fixed (but the number of items and resource capacities increase) and under a suitable probabilistic model of the problem's data. In particular, Romeijn and Romero Morales [95] assume that the values of the costs c_{ij} are continuous and random on a bounded interval $[\underline{C}, \overline{C}]$, the capacity consumption values b_{ij} are continuous and random on the bounded interval $[0, \overline{B}]$ for each $i \in I$ and $j \in J$, and that the resource capacity values K_j increase linearly in the number of items $|I|$ according to the relationship $K_j = \kappa_j|I|$ for some positive constant κ_j and for all $j \in J$. Beyond these assumptions, an additional technical assumption is required in order to ensure asymptotic feasibility with probability one as the number of items $|I|$ increases under the probabilistic model for the input data. Under this technical assumption, Romeijn and Romero Morales demonstrate the asymptotic optimality of the modified greedy heuristic as $|I|$ goes to infinity.

To gain some insight into this result, we need to employ two important properties that we have discussed. The first such property says that if $x_{ij} = 1$ in the LP relaxation solution, then $x_{ij} = 1$ in the modified greedy heuristic solution. The second property states that the number of items that are split between resources is bounded by the number of resources, $|J|$. Therefore, there are at most $|J|$ items whose associated variable values differ between the LP relaxation solution and the modified greedy heuristic solution. Because the optimal solution value will grow to infinity as the number of jobs, $|I|$, goes to infinity, we need to normalize the objective function value by dividing by the number of jobs $|I|$ in order to obtain meaningful asymptotic results. Let Z^{LP} and Z^{MGH} denote the LP relaxation solution value and the modified greedy heuristic solution value, respectively. Then we are interested in the difference between the normalized solution values, i.e., we would like to characterize the behavior of

$$\frac{Z^{MGH}}{|I|} - \frac{Z^{LP}}{|I|} = \frac{Z^{MGH} - Z^{LP}}{|I|}, \tag{4.34}$$

as $|I| \to \infty$ (note that (4.34) must be nonnegative because the optimal LP relaxation value provides a lower bound on any feasible solution value). If x_{ij}^{LP} and x_{ij}^{MGH} equal, respectively, the assignment variable values in the optimal LP relaxation solution and heuristic solution, then $Z^{LP} = \sum_{i \in I} \sum_{j \in J} c_{ij} x_{ij}^{LP}$ and $Z^{MGH} = \sum_{i \in I} \sum_{j \in J} c_{ij} x_{ij}^{MGH}$. Let $S \subseteq I$ denote the set of items with assignments split between resources in the optimal LP relaxation and let $NS = I \backslash S$ denote the complement of S. Recall that $|S| \leq |J|$ because the number of split items in the optimal LP relaxation solution cannot exceed the number of resources. We know that

$$Z^{MGH} - Z^{LP} = \sum_{i \in S} \sum_{j \in J} c_{ij} x_{ij}^{MGH} - \sum_{i \in S} \sum_{j \in J} c_{ij} x_{ij}^{LP}, \tag{4.35}$$

because $\sum_{i \in NS} \sum_{j \in J} c_{ij} x_{ij}^{MGH} = \sum_{i \in NS} \sum_{j \in J} c_{ij} x_{ij}^{LP}$ and $NS \cup S = I$, with

$NS \cap S = \emptyset$. Observe next that

$$\sum_{i \in S} \sum_{j \in J} c_{ij} x_{ij}^{MGH} \leq \overline{C} \sum_{i \in S} \sum_{j \in J} x_{ij}^{MGH} = \overline{C}|S|, \tag{4.36}$$

where the first inequality follows from the bounded interval for c_{ij} values and the equality follows from the assignment constraints (4.3). Using similar logic, we also have that

$$\sum_{i \in S} \sum_{j \in J} c_{ij} x_{ij}^{LP} \geq \underline{C} \sum_{i \in S} \sum_{j \in J} x_{ij}^{LP} = \underline{C}|S|. \tag{4.37}$$

Inequalities (4.36) and (4.37) taken together with (4.35) imply that

$$\frac{Z^{MGH} - Z^{LP}}{|I|} \leq \frac{\left(\overline{C} - \underline{C}\right)|S|}{|I|} \leq \frac{\left(\overline{C} - \underline{C}\right)|J|}{|I|}, \tag{4.38}$$

because $|S| \leq |J|$. Because $\left(\overline{C} - \underline{C}\right)|J|$ is fixed, the right-hand side of (4.38) goes to zero as $|I|$ goes to infinity, which implies that the difference between the normalized LP relaxation solution value and the normalized modified greedy heuristic solution value approaches zero as $|I|$ goes to infinity, with probability one. This implies the asymptotic optimality of the modified greedy heuristic.

Example 4.3 Returning to the problem from Examples 4.1 and 4.2, suppose we apply the modified greedy heuristic described in this section. The optimal dual multipliers from the GAP formulation corresponding to the capacity constraint set (4.2) are given by $\alpha_1 = 8$ and $\alpha_2 = 3.8333$. The entries in the table below show the values of $f_\alpha(i, j)$, while the bottom row contains the value of ρ_i for each item $i = 1, \ldots, 7$.

$f_\alpha(i, j)$ values

Item, i	1	2	3	4	5	6	7
Resource 1	26	25	25	35	18	25	19
Resource 2	26	27	25	7.833	12.667	8.833	7.833
ρ_i	0	2	0	27.167	5.333	16.167	11.167

Using the modified greedy heuristic, we process items in the order 4, 6, 7, 5, 2, and then either item 1 then item 3, or item 3 then item 1, due to the tie between ρ_1 and ρ_3. If we process item 1 before item 3, and assign item 1 to resource 2, then we obtain the optimal GAP solution given in Example 4.1 with objective function value 31. If we process item 3 before item 1, and assign item 3 to resource 2, then we do not obtain a feasible solution. This illustrates the fact that the heuristic does not guarantee finding a solution, and that the order in which ties are broken influences the ability for the heuristic to find a feasible solution. Finally, note that if we start the branch-and-price method with the columns corresponding to the feasible solution given by the modified greedy algorithm, this instance of the SP(GAP) terminates at the root node with an LP relaxation solution value equal to 31. Thus, for this problem instance, the LP relaxation of the SP(GAP) at the root node has zero integrality gap. In addition, the optimal dual multipliers associated with constraint set (4.23) for the LP relaxation of the SP(GAP) formulation also maximize the Lagrangian dual for this problem instance. □

4.6 Review

Although we have not discussed the use of valid inequalities for the GAP, it is easy to see that each of the capacity constraints (4.2) takes the form of a knapsack constraint. As a result, the simple cover inequalities we discussed in Chapter 2 are valid for the GAP when applied to each of the problem's knapsack inequalities. Gottlieb and Rao [59] describe generalizations of these cover inequalities that are valid for the GAP, and provide conditions under which these inequalities form facets of the convex hull of GAP solutions (as well as results for additional classes of inequalities for the GAP). Later, in Section 8.5.2, when discussing valid inequalities for the discrete facility location problem, we present a set of *indexing inequalities*, which are valid for the special case of the GAP in which assignment costs and item capacity consumption values are resource-independent. For additional polyhedral results for the GAP, please see [37] and [38].

This chapter introduced the concept of Lagrangian relaxation and demonstrated its application to the GAP. In addition, we applied a branch-and-price approach for the GAP because of its ability to be formulated as a set partitioning problem. The asymptotically optimal heuristic described in the previous section makes use of important structural properties of the GAP formulation's LP relaxation, wherein the number of items that are split among resources could be bounded by the number of resources. These results imply that a fairly simple heuristic approach can be applied that works very well when the number of items is large and the number of resources is relatively small. Interestingly, we saw that the branch-and-price approach tends to work well when the number of items per resource, $|I|/|J|$, is relatively small. In contrast, the modified greedy heuristic tends to be more effective when the number of items per resource is relatively large. Therefore, taken together, these approaches permit covering a broad variety of problem instances with effective solution approaches.

Exercises

Ex. 4.1 — Given that the multiple knapsack problem is strongly $\mathcal{NP}$-Complete, argue why this implies that the GAP is strongly $\mathcal{NP}$-Complete.

Ex. 4.2 — Show that if $f(x) : \mathbb{R}^n \rightarrow \mathbb{R}$ is a concave function, and $\bar{x}$ does not maximize $f(x)$, then for a sufficiently small step in the direction ξ, where $\xi \in \partial f(\bar{x})$, $f(x)$ increases, and we move closer to an optimal solution for maximizing $f(x)$.

Ex. 4.3 — For the problem in Examples 4.1 and 4.2, show that the dual multipliers given in Example 4.2 lead to a Lagrangian dual solution value (and, therefore, lower bound) equal to the optimal solution of the corresponding instance of the GAP.

Ex. 4.4 — By definition, for a matrix that is totally unimodular, the determinant of every square submatrix equals -1, 1, or 0. Explain why this implies that if we have a constraint matrix that is totally unimodular and a right-hand side vector that is integer, then all extreme points associated with the constraint set are integer extreme points. (Hint: Use Cramer's Rule, which states that for a system of n equations in n unknowns, $Ax = b$, with a unique solution, this solution is given by $x_i = \det(A_i)/\det(A)$, where $\det(\cdot)$ is the determinant of a square matrix, and A_i consists of the A matrix with the ith column replaced by the vector b).

Ex. 4.5 — Suppose that the pricing problem for the set partitioning formulation of the GAP is identical for two different resources, and we therefore obtain the same solution for problem PP_j for these two resources. Explain why it is possible for the resulting column to "price out" (to have an attractive reduced cost) for one resource and not for the other.

5

Uncapacitated Economic Lot Sizing Problems

5.1 Introduction

Perhaps the most widely studied and simplest combinatorial optimization problem in production planning, the uncapacitated economic lot sizing problem (UELSP) was first solved in the 1958 seminal paper by Wagner and Whitin [116]. Sometimes referred to as dynamic lot sizing (or requirements planning), this problem addresses the basic tradeoff between production ordering costs and inventory holding costs for a single stage that faces dynamic and deterministic customer demands. The problem is divided into a contiguous set of discrete planning periods, and the goal is to meet all demands at minimum total production and inventory holding cost over a finite planning horizon consisting of the set of planning periods. In any planning period, an order cost exists for acquiring inventory, independent of the order quantity, and a holding cost is assessed against the inventory held in the period. While we will approach the problem as if placement of an order results in immediate replenishment (i.e., zero replenishment lead time), we can easily account for a positive constant replenishment lead time by offsetting the orders placed in the zero-lead-time case by the constant lead time value.

Solving the UELSP requires specifying the order quantity in each period of the planning horizon, and the sequence of order quantities determines the total production and inventory holding cost incurred. Ordering more frequently results in higher order costs, but reduces inventory holding cost, while infrequent orders reduce order costs at the expense of higher holding costs. This problem therefore addresses the same essential tradeoff as the classic economic order quantity (EOQ) model [65] (discussed in more detail in Chapter 7), but permits a more general set of cost and demand assumptions, as the demand rate and costs must take time-invariant values in the EOQ model.

Applications of the basic UELSP model are abundant in practice, as contexts involving production/procurement economies of scale (e.g., a fixed production setup cost or procurement order cost that is independent of the number of units ordered) fit within this scope. From a practical standpoint, criticisms of the application of this model are seemingly equally abundant, as the model ignores a number of relevant practical factors, including production/procurement capacity limits, demand and lead-time uncertainty, and resource sharing among different products or production stages. These criticisms stem from the use of the model in solving real-world problems for which the model's assumptions are inappropriate (e.g., widespread usage of the model in commercial material requirements planning (MRP) systems applied to contexts violating the model assumptions). A fundamental understanding of this basic model and its assumptions, however, provides insight on applications for which the model is most appropriate. Moreover, we can then often generalize the model to account for the unique features of a given problem context, resulting in a model that more accurately captures the practical reality.

Our goal for this chapter is to provide an intuitive explanation of many of the important results for this problem class since an efficient algorithm was first developed in 1958 [116]. We focus on providing classical dynamic programming solution approaches, and we also highlight key structural and algorithmic results that assist in providing a solid foundation for studying economic lot sizing problems.

5.2 The Basic UELSP Model

Consider a single production stage that requires meeting a sequence of demands over a T-period planning horizon with periods indexed by $t = 1, 2, \ldots, T$, where demand in period t equals d_t (for ease of exposition, we assume all $d_t > 0$; the modifications to handle cases where some d_t values are zero are straightforward and are therefore omitted). We initially assume that shortages are not permitted, although we later relax this restriction. The sequence of events in this model is as follows. At the beginning of each period t the producer determines a nonnegative production quantity x_t, which is produced (or delivered) instantaneously. After order delivery, demand is realized in period t, and inventory cost is assessed against the remaining inventory after demand realization.[1] Assuming zero initial inventory, then given a vector of production quantities $x = (x_1, x_2, \ldots, x_T)$, the inventory remaining at the end of period t, which we denote by i_t is given by the equation $i_t = \sum_{\tau=1}^{t} x_\tau - \sum_{\tau=1}^{t} d_\tau$. In principle we can specify production and inventory holding cost functions in each period, e.g., $\tilde{c}_t(x_t)$ and $\tilde{h}_t(i_t)$. Because of our emphasis on mixed integer programming models, and because this approach is commonly employed in the operations literature, we consider specific functional forms for holding costs that are proportional to the end-of-period inventory and production costs in-

[1]Note that assessing inventory cost based on end-of-period inventory is a commonly employed convention in the literature. Other conventions, such as assessing inventory cost against beginning inventory or against the average between the beginning and ending inventory may alternatively be employed, and we require specifying a particular convention in our model.

volving a fixed-charge structure. Letting h_t denote a (nonnegative) per-unit holding cost in period t, we therefore have $\tilde{h}_t(i_t) = h_t i_t$. For the fixed-charge production cost structure, we let f_t denote the (nonnegative) fixed order cost in period t, while c_t denotes the (nonnegative) variable production cost incurred for each unit ordered. We thus have $\tilde{c}_t(x_t) = f_t y_t + c_t x_t$, where y_t is a binary indicator variable that equals 1 if we order a positive quantity in period t, while $y_t = 0$ if $x_t = 0$. Observe that this production cost function is concave in the production quantity x_t. We can now provide the following MILP formulation for the UELSP.

[UELSP] Minimize $\sum_{t=1}^{T} \{f_t y_t + c_t x_t + h_t i_t\}$ (5.1)

Subject to:

$$i_{t-1} + x_t = d_t + i_t, \quad t = 1, \ldots, T, \tag{5.2}$$

$$x_t \leq M_t y_t, \quad t = 1, \ldots, T, \tag{5.3}$$

$$i_0 = 0, x_t, i_t \geq 0, \quad t = 1, \ldots, T, \tag{5.4}$$

$$y_t \in \{0, 1\}, \quad t = 1, \ldots, T. \tag{5.5}$$

The objective function of the UELSP (5.1) minimizes the total fixed and variable production plus holding costs. Constraints (5.2) ensure that the available amount of the item prior to demand fulfillment (the left-hand side of the constraint) equals the demand plus the end-of-period inventory. The parameter M_t in constraint set (5.3) corresponds to a large positive value, which enables this constraint to force production in period t to equal zero if $y_t = 0$, and effectively leaves the production amount unlimited otherwise. We can, without loss of optimality, set $M_t = \sum_{\tau=t}^{T} d_\tau \equiv D(t, T)$, as an optimal solution will never produce more than this in period t. The final two constraint sets (5.4) and (5.5) specify the variable restrictions.

5.2.1 Fixed-charge network flow interpretation

The inventory balance constraints (5.2) have a network flow structure, and we can think of each period as having a *demand node* as shown in Figure 5.1. The demand node in period t has a requirement of d_t, which is supplied by an inventory arc with flow i_{t-1} and a production arc with flow x_t. An outgoing inventory arc from the demand node in the period carries any excess flow (the total incoming flow, less d_t) to the demand node in period $t + 1$, and the flow on this arc equals i_t. No incoming inventory arc is required for the period 1 demand node, while no outgoing inventory arc is required for the demand node in period T. A source node carrying a supply equal to $D(1, T)$ contains an outgoing arc to each demand node, which completes the specification of the network structure. The resulting network flow problem differs from standard minimum cost network flow (MCNF) problems as a result of the fixed charges associated with flow on the production arcs. We can, however, utilize a number of key properties of MCNF problems to derive important structural properties of optimal solutions for our fixed-charge network flow problem.

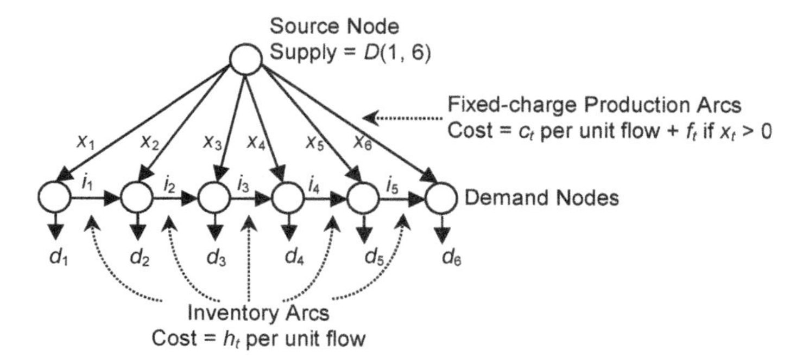

FIGURE 5.1
Fixed-charge network flow representation for a six-period instance of UELSP.

The UELSP minimizes a concave function of the flows on the network shown in Figure 5.1, which implies that an extreme point optimal solution exists.[2] Extreme points of sets defined by network flow constraints induce so-called *spanning trees* on the network, as we next describe. A spanning tree on a connected network corresponds to a connected subgraph containing no (undirected) cycles (therefore, in a spanning tree, a unique undirected path exists between each pair of nodes). From linear programming theory, we know that a one-to-one correspondence exists between extreme point solutions and *basic feasible solutions* (see Footnote 4 on Page 13 for a characterization of basic feasible solutions). Given an extreme point solution, if we draw only the arcs corresponding to the basic variables, the resulting subnetwork will not contain any (undirected) cycles (see Ahuja, Magnanti, and Orlin [6]). Considering the network shown in Figure 5.1, this implies that extreme point solutions for the UELSP never simultaneously have flow into the demand node in period t from both the incoming inventory arc and the incoming production arc, i.e., we never produce in period t if we have inventory remaining at the end of period $t-1$. This is often called the *zero-inventory ordering* (ZIO) property in the literature, and it leads to a very efficient solution approach for the UELSP. This property allows us to consider solutions such that if we produce any positive amount in period t, we will produce an amount exactly equal to the next τ periods of demand for some $\tau = 1, \ldots, T-t+1$, i.e., we need only consider a total of $T-t+1$ production quantities $D(t,t), D(t,t+1), \ldots, D(t,T)$.

[2]Given a concave function $f(x)$ that we wish to minimize over a polyhedron with extreme points $x^1, x^2, \ldots, x^K$, suppose we have an optimal point x' that is not an extreme point. Then x' can be written as a convex combination of the extreme points, i.e., $x' = \sum_{k=1}^{K} \lambda_k x^k$ with all $0 \leq \lambda_k \leq 1$ and $\sum_{k=1}^{K} \lambda_k = 1$. Concavity of $f(x)$ implies $f(x') \geq \sum_{k=1}^{K} \lambda_k f(x^k)$. This implies that either an extreme point exists with a lower objective function value than x' (a contradiction to the optimality of x') or multiple extreme points exist with the same objective function value as x', which confirms the claim.

The following section utilizes this property to provide an efficient dynamic programming algorithm for the UELSP.

5.2.2 Dynamic programming solution method

The ZIO property implies that in our search for an optimal solution, we only need to consider $T - t + 1$ values for the production quantity in any period t. We can characterize such solutions completely by specifying the periods in which orders are placed. A solution containing n orders, for example, can be specified by the vector of order periods $\mathbb{T} = (t_1, t_2, \ldots, t_{n+1})$, where $1 = t_1 < t_2 < t_3 < \ldots < t_{n+1} = T+1$. The cost of this solution, $C(\mathbb{T})$, can be computed as

$$C(\mathbb{T}) = \sum_{i=1}^{n} g(t_i, t_{i+1}), \tag{5.6}$$

where $g(t_i, t_{i+1}) = f_{t_i} + c_{t_i} D(t_i, t_{i+1} - 1) + \sum_{j=1}^{t_{i+1}-t_i-1} h_{t_i+j-1} D(t_i + j, t_{i+1} - 1)$ is the cost of ordering $D(t_i, t_{i+1} - 1)$ units in period t_i in order to satisfy all demand in periods $t_i, \ldots, t_{i+1} - 1$. Each pair (t_i, t_{i+1}) defines a *regeneration interval* (RI), which is characterized by a pair of periods (s, u) with $u > s$ such that $i_{s-1} = 0$, $i_{u-1} = 0$, and $i_t > 0$ for $t = s, s+1, \ldots, u-2$. We will have more to say about regeneration intervals when discussing the backordering case in Section 5.5, as well as the capacitated economic lot sizing problem (CELSP) in the next chapter. For now we note that specifying a sequence of order periods for the UELSP is equivalent to defining a sequence of RIs, and that the cost of an RI is independent of the sequence of RIs that precede or follow it. Using Equation (5.6) directly to determine an optimal solution apparently requires the evaluation of 2^{n-1} values of the vector $\mathbb{T}$. Enumerating each of these possible vector values is not necessary, however, because we can reduce the number of required computations to a manageable value by using dynamic programming.

We define $G(t)$ as the minimum cost of a production plan for periods t through T. Defining $G(T+1) = 0$, we can compute $G(t)$ for $1 \leq t \leq T$ recursively using

$$G(t) = \min_{\tau = t+1, \ldots, T+1} \{g(t, \tau) + G(\tau)\}, \tag{5.7}$$

with an optimal UELSP solution value given by $G(1)$. When computing $G(t)$ for a given value of t, let $\tau^*(t)$ denote the value of τ that provides the minimum in (5.7), where $\tau^*(t)$ provides the next order period following period t in an optimal solution for the problem from period t to T ($\tau^*(t) = T + 1$ implies that period t is the final order period in the solution). To recover the optimal set of order periods for the UELSP, we proceed forward beginning with $\tau^*(1)$, and stopping when $\tau^*(t) = T + 1$ for some t encountered in the process. Using the recursion (5.7) requires computing $\tau^*(t)$ (which requires $\mathcal{O}(T)$ operations for a fixed value of t) for each of the T values of t, which implies that the UELSP can be solved using this dynamic programming procedure in $\mathcal{O}(T^2)$ time.

We can illustrate a convenient and intuitive representation of this solution method using a shortest path graph, the structure of which is quite different from the network shown in Figure 5.1. For a T-period problem, we construct a graph containing $T+1$ nodes, where the nodes are numbered according to time periods. Given a node, we create an arc leaving the node and entering every higher numbered node. The cost of arc (s, u) for any $1 \leq s < u \leq T+1$ then equals $g(s, u)$. Any path in the graph corresponds to a sequence of orders, and for every sequence of orders, a path exists in the graph from node 1 to node $T+1$, with a cost equal to the corresponding production plan cost. Finding the shortest path in this graph then provides the minimum cost production plan, with the nodes encountered along the path indicating the periods in which orders are placed. Figure 5.2 illustrates the associated network structure for a five-period problem instance.

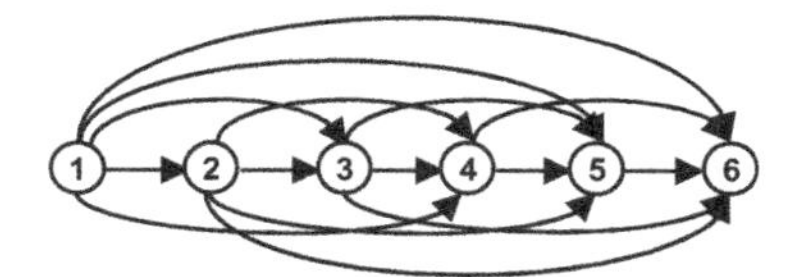

The cost of arc (i, j) equals the fixed order cost in period i plus the variable production cost $c_iD(i, j-1)$, plus the holding cost incurred in meeting demand in periods $i, \ldots, j-1$, using production in period i.

FIGURE 5.2
Shortest path network structure for a five-period instance of UELSP.

5.3 Tight Reformulation of UELSP

The UELSP formulation in Section 5.2 is easily explained and interpreted, particularly in terms of the fixed-charge network flow interpretation. This formulation has its drawbacks, however, for a number of reasons. In particular, when using this formulation, the LP relaxation solution value (obtained by relaxing the integrality restrictions on the y_t variables) is often strictly less than the optimal MILP solution value. In this section we provide a reformulation of the problem and show that the optimal LP relaxation value for this new formulation is equal to the optimal MILP solution value.

Recall that we can write $i_t = \sum_{\tau=1}^{t} x_\tau - \sum_{\tau=1}^{t} d_\tau$, and there is, therefore, no need for including these inventory variables in our formulation. Making this substitution throughout the UELSP formulation in Section 5.2, and defining $\hat{c}_t = c_t + \sum_{\tau=t}^{T} h_\tau$, results in the following equivalent formulation, UELSP$'$.

$$
\textbf{[UELSP}'\textbf{]} \quad \text{Minimize} \quad \sum_{t=1}^{T} \{f_t y_t + \hat{c}_t x_t\} \tag{5.8}
$$

$$
\text{Subject to:} \quad \sum_{\tau=1}^{t} x_\tau \geq \sum_{\tau=1}^{t} d_\tau, \qquad t = 1, \ldots, T, \tag{5.9}
$$

$$
x_t \leq D(t,T) y_t, \qquad t = 1, \ldots, T, \tag{5.10}
$$

$$
x_t \geq 0, \qquad t = 1, \ldots, T, \tag{5.11}
$$

$$
y_t \in \{0,1\}, \qquad t = 1, \ldots, T. \tag{5.12}
$$

Note that we have omitted the constant term $-\sum_{t=1}^{T} d_t \sum_{\tau=t}^{T} h_\tau$ from the objective function without loss of optimality. This reformulation illustrates the fact that for any instance of the UELSP, an equivalent problem exists with zero holding costs and a variable production cost of $\hat{c}_t$ in period t. We next make an additional set of substitutions to obtain the tight formulation that we will analyze in greater detail. With a slight abuse of notation, define $x_{t,\tau}$ as the *proportion* of period τ demand that we satisfy using production in period t, and observe that $x_t = \sum_{\tau=t}^{T} d_\tau x_{t,\tau}$. We also define $\hat{c}_{t,\tau} = \hat{c}_t d_\tau$ to obtain the following tight reformulation, UELSP$''$.

$$
\textbf{[UELSP}''\textbf{]} \quad \text{Minimize} \quad \sum_{t=1}^{T} f_t y_t + \sum_{t=1}^{T} \sum_{\tau=t}^{T} \hat{c}_{t,\tau} x_{t,\tau} \tag{5.13}
$$

$$
\text{Subject to:} \quad \sum_{t=1}^{\tau} x_{t,\tau} = 1, \qquad \tau = 1, \ldots, T, \tag{5.14}
$$

$$
0 \leq x_{t,\tau} \leq y_t, \qquad t = 1, \ldots, T, \; \tau = t, \ldots, T, \tag{5.15}
$$

$$
y_t \in \{0,1\}, \qquad t = 1, \ldots, T. \tag{5.16}
$$

As we will later see when we discuss location problems in Chapter 8, formulation UELSP$''$ is a structured special case of the uncapacitated discrete facility location problem (UFLP). Although the general UFLP is an $\mathcal{NP}$-Hard optimization problem, the special structure of the UELSP permits polynomial time solution, as we saw in the previous section. The LP relaxation of UELSP$''$ is obtained by replacing constraints (5.16) with $0 \leq y_t \leq 1$ for all t. To show that this LP relaxation has the same optimal solution value as that of formulation UELSP$''$, we use a proof technique that applies Lagrangian relaxation.

5.3.1 Lagrangian relaxation shows a tight formulation

Because the Lagrangian dual problem provides a lower bound on the optimal solution value of the original problem, if the solution to the Lagrangian dual is also feasible for the original problem, then it must also be optimal for the original problem. In particular, if we apply Lagrangian relaxation to formulation UELSP″ and the optimal Lagrangian dual solution is feasible for UELSP″, then we can claim that this solution is optimal for UELSP″. Moreover, if the resulting Lagrangian relaxation problem satisfies the integrality property, then the optimal LP relaxation solution value will equal the optimal Lagrangian dual objective value. If both of these solution values equal the optimal solution value of UELSP″ then we have our result (i.e., that the LP relaxation is tight, with $Z^*_{LP} = Z^*_{IP}$).

Returning to formulation UELSP″, suppose we relax the constraints $x_{t,\tau} \leq y_t$, $t = 1, \ldots, T$ and $\tau = t, \ldots, T$, using the nonnegative Lagrangian multipliers $\lambda_{t,\tau}$, $t = 1, \ldots, T$ and $\tau = t, \ldots, T$, which introduces the term $\sum_{t=1}^{T}\sum_{\tau=t}^{T} \lambda_{t,\tau}(x_{t,\tau} - y_t)$ to the objective function, and eliminates these constraints from the formulation. The resulting Lagrangian relaxation formulation becomes

[LR(UELSP)]

$$Z_{LR}(\lambda) = \text{Min} \sum_{t=1}^{T} \left(f_t - \sum_{\tau=t}^{T} \lambda_{t,\tau} \right) y_t + \sum_{t=1}^{T}\sum_{\tau=t}^{T} (\hat{c}_{t,\tau} + \lambda_{t,\tau})\, x_{t,\tau} \quad (5.17)$$

$$\text{Subject to:} \quad \sum_{t=1}^{\tau} x_{t,\tau} = 1, \quad \tau = 1, \ldots, T, \quad (5.18)$$

$$x_{t,\tau} \geq 0, \quad t = 1, \ldots, T, \quad \tau = t, \ldots, T, \quad (5.19)$$

$$y_t \in \{0, 1\}, \quad t = 1, \ldots, T. \quad (5.20)$$

Clearly the linear relaxation of the feasible region of LR(UELSP) has all integer extreme points and therefore satisfies the integrality property, which implies that the optimal LP relaxation solution value will equal the optimal solution value of the Lagrangian dual problem. We can determine $Z_{LR}(\lambda)$ given a vector $\lambda \geq 0$ as follows. If $f_t - \sum_{\tau=t}^{T} \lambda_{t,\tau} \leq 0$ set $y_t = 1$; otherwise, set $y_t = 0$. Let $\mathcal{T}(\lambda)$ denote the set of periods t such that $y_t = 1$. For $\tau = 1, \ldots, T$, let $\Theta(\tau, \lambda)$ denote the set of all $t \leq \tau$ such that $\hat{c}_{t,\tau} + \lambda_{t,\tau}$ takes a minimum value. For each $\tau = 1, \ldots, T$, if some $t \leq \tau$ exists such that $t \in \{\mathcal{T}(\lambda) \cap \Theta(\tau, \lambda)\}$ (i.e., such that $y_t = 1$ and $\hat{c}_{t,\tau} + \lambda_{t,\tau}$ is minimized), then set $x_{t,\tau} = 1$; otherwise, choose any $t \in \Theta(\tau, \lambda)$ and set $x_{t,\tau} = 1$. Observe that the resulting solution provides $Z_{LR}(\lambda)$.

Let λ^* denote a vector of Lagrangian multipliers that maximizes the Lagrangian dual, and let $x^*(\lambda^*), y^*(\lambda^*)$ denote the corresponding Lagrangian relaxation problem solution. If $x^*(\lambda^*), y^*(\lambda^*)$ is feasible for UELSP″, then we have the desired result, i.e., $Z^*_{LP} = Z_{LD} = Z^*_{IP}$. Otherwise, there must exist some $(\tilde{t}, \tilde{\tau})$ pair such that $x_{\tilde{t},\tilde{\tau}} = 1$ with $y_{\tilde{t}} = 0$, and we know by construction that $y_t = 0$ for all $t \in \Theta(\tilde{\tau}, \lambda^*)$. We can obtain a new vector $\tilde{\lambda}$ of multipliers by increasing $\lambda^*_{t,\tilde{\tau}}$ for each $t \in \Theta(\tilde{\tau}, \lambda^*)$ by some $\epsilon > 0$, and leaving the remaining values of $\lambda^*_{\tau,t}$ unchanged. For ϵ sufficiently small, the coefficient of y_t for all $t \in \Theta(\tilde{\tau}, \tilde{\lambda})$ remains nonnegative and the set $\Theta(\tilde{\tau}, \tilde{\lambda})$ is identical to $\Theta(\tilde{\tau}, \lambda^*)$. Thus, $x^*(\lambda^*), y^*(\lambda^*)$ is an optimal solution for $Z_{LR}(\tilde{\lambda})$ with

$Z_{LR}(\tilde{\lambda}) > Z_{LR}(\lambda^*)$, contradicting the optimality of λ^*. Thus, no such violation of feasibility exists for UELSP$''$ at the optimal dual vector λ^*, and we have the desired result. That is, an optimal solution exists for the Lagrangian dual that is feasible for the original integer program, i.e., $Z_{LD} = Z_{IP}^*$, which also implies that this point solves the LP relaxation.

Example 5.1 Consider a five-period instance of the UELSP with the parameters shown in the table below (assume for this example that variable production costs are the same in every period; thus, total variable production costs are constant and may be ignored).

Period, t	1	2	3	4	5
f_t	110	95	115	95	110
h_t	0.90	1.10	0.95	1.00	1.00
$\hat{c}_t$	4.95	4.05	2.95	2.00	1.00
d_t	50	83	28	68	33
M_t	262	212	129	101	33

Using the UELSP formulation with the M_t values shown in the table, the LP relaxation solution value equals 211.15, while the tight formulation (and the optimal MILP) solution value equals 363.80. Thus, the UELSP formulation has a relatively large integrality gap of $100\% \times (363.80 - 211.15)/363.80 \approx 41.96\%$. We will return to this example in the next section. □

5.4 An $\mathcal{O}(T \log T)$ Algorithm for the UELSP

In the dynamic programming recursion (5.7), as our previous reformulation exercises indicate, we can equivalently replace the term $g(t, \tau)$ with $f_t + \hat{c}_t D(t, \tau - 1)$. The fixed cost term f_t is incurred regardless of the value of τ, and we can solve the following problem to determine the value of τ that provides $G(t)$, the minimum cost solution value for periods t through T:

$$\hat{G}(t) = \min_{\tau = t+1, \ldots, T+1} \{\hat{c}_t D(t, \tau - 1) + G(\tau)\}. \tag{5.21}$$

Observe that $G(t) = \hat{G}(t) + f_t$. As we noted previously, direct computation of $G(t)$ (by computing the value in the brackets in (5.21) for each $\tau > t$) requires $\mathcal{O}(T)$ operations. Given values of t and $G(\tau)$ for all $\tau > t$, however, it is possible to determine the minimizing value of τ in $\mathcal{O}(\log T)$ operations, which implies an overall computational effort of $\mathcal{O}(T \log T)$ for solving the UELSP. Wagelmans, van Hoesel, and Kolen [115] illustrated how to see this in an intuitive way. In solving for $\hat{G}(t)$ (which equivalently determines $G(t)$), we create a graph as shown in Figure 5.3 with a "cumulative demand satisfied axis" (the abscissa, denoted by D in the figure) and a "cost-to-go axis" (the ordinate, denoted by G in the figure), and plot the pairs $(D(t, \tau-1), G(\tau))$, for $\tau = t+1, \ldots, T+1$, ensuring that the abscissa extends out to the value $D = D(t,T)$. Evaluating the term in the brackets in (5.21) for a given τ is equivalent to projecting a line with slope $-\hat{c}_t$ from the point $(D(t, \tau-1), G(\tau))$ backwards to the point where this line intersects the vertical axis. The corresponding cost at the point of intersection with the vertical axis provides the minimum cost to go from period t (less the fixed order cost in period t, f_t) given that the next order following period t occurs in period τ; the minimum value among these intersection points determines $\hat{G}(t)$.

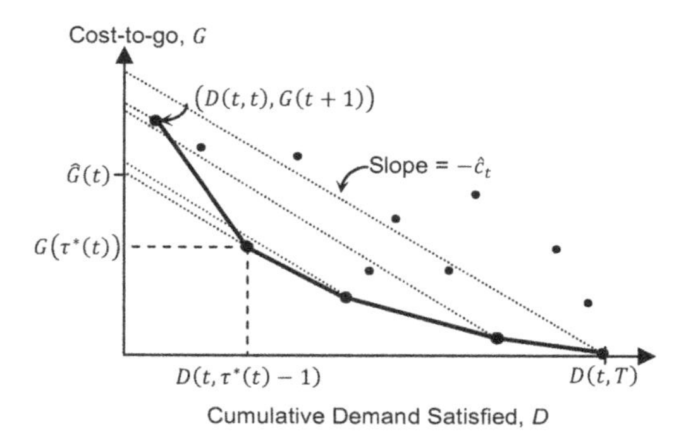

FIGURE 5.3
Cost-to-go versus demand-to-go graph to illustrate minimization approach for determining $\hat{G}(t)$ (5.21) (figure adapted from Figure 3 in [115]).

Our prior approach in Section 5.2.2 essentially evaluates this term for all $\tau > t$ and for each t. By considering this graphical representation of the minimization problem in (5.21), however, it is clear that we need not explicitly evaluate this term for every higher numbered period than t. In particular, if the pair $(D(t, \tau-1), G(\tau))$ does not lie on the lower convex envelope[3] (LCE) of points on the graph (see Figure 5.3), then we need not consider period τ as a candidate for finding $\hat{G}(t)$, because such points will be dominated by points that do lie on the LCE. Moreover, by considering the graphical representation of this problem, it should be clear that in searching for $\hat{G}(t)$, we are seeking a point $(D(t, \tau-1), G(\tau))$ that lies on the LCE such that a line going through the point with slope $-\hat{c}_t$ supports the epigraph[4] of the LCE (that is, this line lies on or below the LCE at all points). Thus if we can maintain a representation of the LCE (defined by the points on the LCE, which also imply the slopes of line segments connecting these points), we can use binary search to determine such a point, as this point will have the property that $-\hat{c}_t$ lies between the slopes of the LCE on either side of the point. Such a point provides the value

[3]The lower convex envelope (LCE) consists of the set of line segments connecting pairs of points such that no point lies below a line segment.

[4]The epigraph of a function $f(x)$ defined for all $x \in \mathbb{R}^1_+$ is the set of all points (x, z) such that $x \geq 0$ and $z \geq f(x)$.

of τ that defines $\hat{G}(t)$ (and therefore $G(t) = \hat{G}(t) + f_t$) at the intersection of the extended line of slope $-\hat{c}_t$ with the vertical axis, and we denote this value of τ by $\tau^*(t)$ in Figure 5.3. The corresponding binary search among slope values of the LCE can be performed in $\mathcal{O}(\log T)$ time by maintaining an ordered list of the slopes associated with the LCE, and searching this list to determine the two slopes between which $-\hat{c}_t$ lies.

After determining $\hat{G}(t)$ (and thus $G(t)$), we wish to determine $G(t-1)$, which requires the LCE for the problem from period t onward, i.e., we need to update the LCE to account for the point $(D(t-1, t-1), G(t))$. We can easily update the set of points on the LCE after determining $G(t)$ as follows. Define the set of periods corresponding to points on the LCE prior to determining $G(t)$ as *efficient periods* (if the point $(D(t, \tau - 1), G(\tau))$ is on the LCE, then period τ is an efficient period). Add the point $(0, G(t))$ to the graph shown in Figure 5.3 and find the smallest efficient period $\hat{\tau}$ such that the slope of the line connecting $(0, G(t))$ to $(D(t, \hat{\tau} - 1), G(\hat{\tau}))$ is less than the slope of the line segment on the LCE connecting $(D(t, \hat{\tau} - 1), G(\hat{\tau}))$ to the next higher numbered efficient period. All periods greater than or equal to $\hat{\tau}$ that were previously on the LCE will remain there, while all periods less than $\hat{\tau}$ that were previously on the LCE will now be removed. The new point $(0, G(t))$ is then added to the LCE, and it connects to the point $(D(t, \hat{\tau} - 1), G(\hat{\tau}))$ to define a new segment of the LCE. Note that we associate two slope values with each breakpoint of the LCE, the slope to the left of the point and the slope to the right. Therefore, when searching for a point, we simply need to search the ordered sequence of slopes associated with the LCE. When we move back by one period, to consider period $t - 1$, we shift the entire graph to the right by d_{t-1} and repeat the procedure until obtaining $G(1)$. Proceeding in this way, by updating the LCE using binary search at each of the T iterations, we can solve the UELSP in $\mathcal{O}(T \log T)$ operations.

Example 5.2 We return to the problem from Example 5.1 to illustrate the algorithm described in this section. Beginning with period 5, clearly $G(5) = 143$, the cost to satisfy demand in period 5 using production in period 5. To find $G(4)$, we create a graph of the lower convex envelope of efficient periods for $t > 4$. The graph for $t = 4$ has only two points, $(D(4,4), G(5)) = (68, 143)$ and $(D(4,5), 0) = (101, 0)$ (see Figure 5.4), and the LCE consists of the line segment connecting these points (with slope $m = -4.33$). At this point, period 5 is considered to be the only efficient period.

Because $-\hat{c}_4 = -2$ is greater than this slope, $\hat{G}(4)$ takes a value equal to the projection of a line with slope $-\hat{c}_4 = -2$ from the point $(101, 0)$ back to the vertical axis, which gives $\hat{G}(4) = 202$. Adding f_4 to this gives $G(4) = 297$. The slope of the line connecting $(0, 297)$ with the point $(68, 143)$ equals -2.26, which is greater than the slope of the segment to the right of the point $(68, 143)$, which implies that the new LCE will eliminate the line segment to the right of this point (which is the only line segment on the LCE at this point). As a result, period 5 is no longer an efficient period, and the optimal

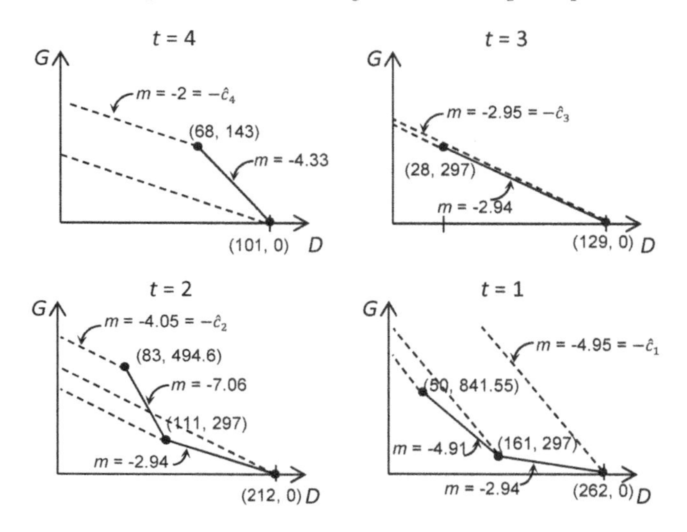

FIGURE 5.4
Illustration of application of $\mathcal{O}(T \log T)$ algorithm to the problem in Example 5.1 (note that m denotes a slope value).

solution from period 4 onward uses the setup in period 4 to satisfy demand in periods 4 and 5, and has cost 297 (period 4 is now the only efficient period).

We next move back to $t = 3$. The LCE for $t = 3$ has one line segment connecting the points $(D(3,3), G(4)) = (28, 297)$ and $(D(3,5), 0) = (129, 0)$, and the slope of this segment equals -2.94. Because $-\hat{c}_3 = -2.95$ is less than this slope, $\hat{G}(3)$ equals the projection of a line with slope $-\hat{c}_3 = -2.95$ from the point $(28, 297)$ back to the vertical axis, which gives $\hat{G}(3) = 379.6$ and $G(3) = 494.6$. The line segment connecting the point $(0, 494.6)$ with $(28, 297)$ has slope -7.06, which is less than the slope of the LCE to the right of the point $(28, 297)$. Thus, no points are eliminated from the LCE, and period 3 is added to the set of efficient periods. The optimal solution from period 3 onward uses production in period 3 to satisfy demand in period 3 and production in period 4 to satisfy demand in periods 4 and 5.

Moving back to period $t = 2$, the LCE now contains the points $(D(2,2), G(3)) = (83, 494.6)$, $(D(2,3), G(4)) = (111, 297)$, and $(D(2,5), 0) = (212, 0)$, and contains two segments with slopes $m = -7.06$ and $m = -2.94$. Because $-\hat{c}_2 = -4.05$ falls between these slopes, $\hat{G}(2)$ is obtained by projecting a line with slope $-\hat{c}_2 = -4.05$ backward to the vertical axis from the point $(111, 297)$, which gives a value of $\hat{G}(2) = 746.55$, implying that $G(2) = 841.55$. The slope of the line connecting the points $(0, 841.55)$ and $(111, 297)$ equals -4.91, which falls between the slopes of the LCE line segments on either side of this point. Thus, period 3 is no longer an efficient period, and the line segment to the left of the point $(111, 297)$ will be replaced on the LCE. The optimal solution from period 2 onward uses production in period 2 to satisfy demand in periods 2 and 3, production in period 4 to satisfy demand in periods 4 and 5, and has a cost of 841.55 (periods 2 and 4 form the set of efficient periods).

Finally, we consider $t = 1$, for which the LCE consists of the line segments joining the points $(D(1,1), G(2))$, $(D(1,3), G(4))$, and $(D(1,5), 0)$, with slope values $m = -4.91$ and $m = -2.94$. Because $-\hat{c}_1 = -4.95$ is smaller than both of these, $\hat{G}(1)$ is determined by projecting a line with slope $-\hat{c}_1 = -4.95$ back to the vertical axis from the point $(D(1,1), G(2)) = (50, 841.55)$, which gives $\hat{G}(1) = 1089.05$ and $G(1) = 1199.05$. We do not eliminate any existing efficient periods, and we add period 1 to the set of efficient periods, which now contains periods 1, 2, and 4. The optimal solution from period 1 forward uses production in period 1 to satisfy period 1 demand, production in period 2 to satisfy demand in periods 2 and 3, and production in period 4 to satisfy demand in periods 4 and 5. The cost of this solution equals 1199.05. Note that the constant term omitted from the objective function in the formulation of UELSP′ (when defining our $\hat{c}_t$ values) equals $-\sum_{t=1}^{T} d_t \sum_{\tau=t}^{T} h_\tau$, which in this problem instance equals 835.25; when this constant is subtracted from 1199.05, this gives an objective function value of 363.8, the optimal solution value given in Example 5.1. □

5.5 Implications of Backordering

Our discussion of the UELSP so far has assumed that shortages are not permitted. We can generalize many of our results to situations in which backordering is permitted at a cost per unit backordered at the end of a period. Allowing backordering is equivalent to permitting negative inventory variable values, although the unit cost of positive inventory (the holding cost) may differ from the unit cost of negative inventory (the backordering cost). Because of this, and because commercial mixed integer programming solvers require nonnegative variables, we redefine our previous inventory variables (i_t) using the superscript (+) to indicate that positive values of these variables imply positive inventory. That is i_t^+ now denotes the amount of inventory on hand, while i_t^- denotes the total amount backordered at the end of period t. This is equivalent to making the substitution $i_t = i_t^+ - i_t^-$ in our original UELSP formulation in Section 5.2, where the i_t variables are unrestricted in sign. In addition, if each unit backordered at the end of period t incurs a shortage cost of b_t, the objective function term $\sum_{t=1}^{T} h_t i_t$ now becomes $\sum_{t=1}^{T} \left\{h_t i_t^+ + b_t i_t^-\right\}$. For completeness, we present the new problem formulation, which we call UELSP(B), below.

[UELSP(B)]

$$\text{Minimize} \quad \sum_{t=1}^{T} \left\{ f_t y_t + c_t x_t + h_t i_t^+ + b_t i_t^- \right\} \tag{5.22}$$

$$\text{Subject to:} \quad i_{t-1}^+ + i_t^- + x_t = d_t + i_t^+ + i_{t-1}^-, \quad t = 1, \ldots, T, \tag{5.23}$$

$$x_t \le M y_t, \quad t = 1, \ldots, T, \tag{5.24}$$

$$i_0^+ = i_0^- = 0, x_t, i_t^+, i_t^- \ge 0, \quad t = 1, \ldots, T, \tag{5.25}$$

$$y_t \in \{0, 1\}, \quad t = 1, \ldots, T. \tag{5.26}$$

Because any period can potentially satisfy all demand over the planning horizon, we can define the big-M value as $M = D(1, T)$. Given any choice of the vector of binary order variables y, the resulting problem is a linear program. Because the variables i_t^+ and i_t^- have linearly dependent columns in this linear program, an optimal solution exists such that $i_t^+ i_t^- = 0$, i.e., we will never have on-hand inventory and outstanding backorders at the end of a period. The fixed-charge network flow representation in Figure 5.5 also illustrates that any solution violating this condition will result a non-extreme point solution, as such a solution will create a cycle in the corresponding subnetwork containing arcs with positive flow. This figure also illustrates an additional property that strengthens our previous ZIO property. That is, in addition to the ZIO property (which requires $x_t i_{t-1}^+ = 0$), for a spanning tree to be induced by the arcs corresponding to basic variables, we must have $i_{t-1}^+ i_t^- = 0$. This implies that we never have a period in which inventory was held from a prior period and backorders exist at the end of the period. Our previous dynamic

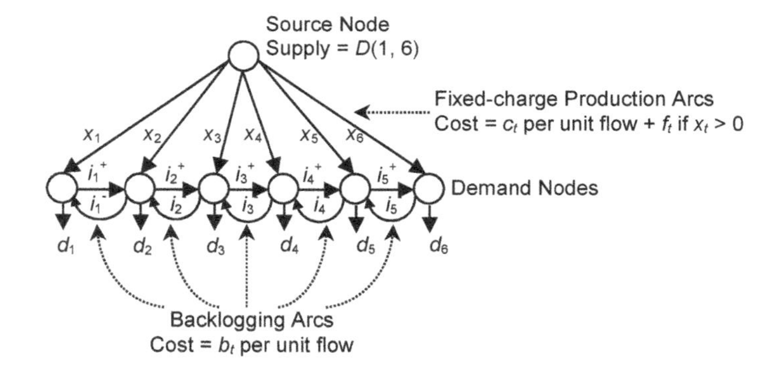

FIGURE 5.5
Fixed-charge network flow representation for a six-period instance of UELSP with backordering permitted.

programming approach exploited the fact that an optimal solution exists such that the production quantity in period t takes a value equal to $D(t, \tau)$ for some $\tau \ge t$. Our new properties for the backordering case now imply that an optimal solution exists such that the production quantity in period t takes a value equal to $D(s,u)$ for some (s,u) values such that $s \leq t \leq u$.

We now slightly alter our definition of a *regeneration interval* (RI) as a pair of periods (s,u) with $u > s$ such that $i_{s-1} = 0$, $i_{u-1} = 0$, and $i_t \neq 0$ for $t = s, s+1, \ldots, u-2$. Given the properties identified in our discussion of the fixed-charge network representation above, an optimal solution exists for the UELSP(B) consisting of a sequence of RIs, where the demand of the RI is supplied by some order period within the RI. That is, for the RI (s,u), an order period t with $s \leq t < u$ supplies the demand $D(s, u-1)$. The order period for the RI (s,u) is determined by solving

$$g(s,u) = \min_{\tau=s,s+1,\ldots,u-1} \left\{ f_\tau + \sum_{i=\tau}^{u-2} h_i D(i+1, u-1) + \sum_{i=s}^{\tau-1} b_i D(s,i) \right\}, \quad (5.27)$$

where any summation equals zero if the upper summation limit is less than the lower summation limit. Observe that computing $g(s,u)$ for any RI (s,u) requires $\mathcal{O}(T)$ operations. If we compute $g(s,u)$ for each of the $\mathcal{O}(T^2)$ RIs, then we can again use our recursive equation (5.7) to solve the problem via dynamic programming, with a total complexity of $\mathcal{O}(T^3)$. Note that this is equivalent to solving the subproblem (5.27) to determine each of the arc costs for the shortest path network shown in Figure 5.2, and then finding the shortest path in this network.

5.6 Review

We confined ourselves to studying the UELSP under a fixed-charge production cost structure. While this leads to the minimization of a concave cost function, the approaches we have provided, for the most part, can be generalized to handle general concave production, inventory, and backordering cost functions. For example, all of the properties that resulted from analysis of the optimality of a spanning tree structure continue to hold under general concave cost functions. Only the results of Sections 5.3.1 and 5.4 explicitly require the fixed plus linear cost structure assumptions, although the computations involved in our dynamic programming solutions require modification when applied to the general concave cost case. The basic approach for solving the UELSP using dynamic programming comes from the seminal paper by Wagner and Whitin [116], while our concave cost network flow representation and interpretation can be traced to Zangwill [117]. The $\mathcal{O}(T \log T)$ algorithm and graphical interpretation were developed by Wagelmans et al. [115]. We note that at about the same time, Aggarwal and Park [5] and Federgruen and Tzur [43] independently published $\mathcal{O}(T \log T)$ algorithms for the UELSP. Modifications of each of the basic algorithms in [5], [43], and [115] also lead to algorithms for solving the model with backordering (in Section 5.5) in $\mathcal{O}(T \log T)$ time (see [115, 110]). Moreover, these algorithms can solve special cases of the problem in $\mathcal{O}(T)$ time. These special cases include instances in which no speculative motives exist for holding inventory (i.e., when the increase in unit production cost does not exceed the unit holding cost from period to period) or when fixed order costs do not decrease with time. Analysis of the backordering case in Section 5.5 can be traced to Zangwill [117].

Exercises

Ex. 5.1 — Show that the zero-inventory ordering property holds without using the spanning tree property of extreme points of network flow problems.

Ex. 5.2 — Consider the UELSP with $f_t = 0$ for all $t = 1, \ldots, T$. Provide an algorithm for solving this problem and characterize the running time of your algorithm.

Ex. 5.3 — For the shortest path solution method illustrated in Figure 5.2, assuming you have computed all of the arc lengths, formulate the problem of finding the shortest path as an optimization problem.

Ex. 5.4 — For the dynamic programming solution method discussed in Section 5.2.2, using Equations (5.6) and (5.7), formulate the problem of minimizing $G(1)$ via a linear program (Hint: the objective should maximize a dummy variable less than or equal to $G(1)$). How many constraints does the resulting formulation have?

Ex. 5.5 — Suppose we use the approach illustrated in Figure 5.3 to solve the problem with zero setup cost in every period. Explain what the lower convex envelope (LCE) will look like at an arbitrary time period if $\hat{c}_t > \hat{c}_{t+1}$ for $t = 1, \ldots, T-1$. What if $\hat{c}_t = \hat{c}_{t+1}$ for $t = 1, \ldots, T-1$?

6

Capacitated Lot Sizing Problems

6.1 Introduction

The previous chapter considered the economic lot sizing problem when no limit exists on the production quantity in any time period. The absence of production capacities permitted demonstrating the optimality of a solution satisfying the so-called zero-inventory ordering (ZIO) property, which eliminates the possibility of production occurring in a period in which inventory was held from the prior period. This, in turn, enabled fully characterizing a solution based solely on the production periods, i.e., the periods in which production is positive. Equally importantly, this property permitted characterizing an optimal solution as a sequence of independent regeneration intervals (RIs), where an RI is defined by a pair of successive production periods (s, u) with $u > s$ such that $i_{s-1} = 0$, $i_{u-1} = 0$, and $i_t \neq 0$ for $t = s, s+1, \ldots, u-2$, where i_t is the inventory level at the end of period t. This definition of an RI combined with the ZIO property implies that within an RI, in the absence of backordering, production only occurs in the first period of the RI; for example, for the RI defined by the period pair (s, u), the production quantity in period s must equal the sum of the demands in periods s through $u - 1$. Otherwise, the ZIO property would be violated.

In the presence of finite production quantity limits, or capacities, it is not unlikely that an optimal solution will violate the ZIO property. To see this, consider a simple two-period problem with a production capacity of k in each period, and suppose that the demand in period 1 equals $k-1$ while the demand in period 2 equals $k + 1$. Assuming that shortages are not permitted, the only feasible solution that meets all demand produces k units in each period, while holding one unit of inventory at the end of period 1. While the ZIO property is no longer useful for the capacity-constrained version of the lot sizing problem, the concept of RIs, as we will see, tends to be very useful in characterizing the properties of optimal solutions. As we will also see, the addition of finite production capacities, in general, leads to a class of economic lot sizing problems that is no longer polynomially solvable (unless $\mathcal{P} = \mathcal{NP}$). This chapter considers this class of $\mathcal{NP}$-Hard capacitated lot sizing problems, discussing properties of optimal solutions for this problem class, a polynomially-solvable practical special case, as well as effective solution approaches that have been developed for the general version of the problem. Except where specifically noted, we assume that shortages are not permitted.

6.2 Capacitated Lot Sizing Formulation

The capacitated version of the economic lot sizing problem (CELSP) is a very simple generalization of the UELSP formulation in the previous chapter. We employ the notation defined in the previous chapter, with the addition of a production capacity limit in period t, denoted by k_t, for $t = 1, \ldots, T$. The MILP formulation of the CELSP is nearly identical to that of the UELSP, with the exception that each big-M_t value is replaced by the corresponding capacity value k_t:

[CELSP]

$$\text{Minimize} \quad \sum_{t=1}^{T} \{f_t y_t + c_t x_t + h_t i_t\} \tag{6.1}$$

$$\text{Subject to:} \quad i_{t-1} + x_t = d_t + i_t, \quad t = 1, \ldots, T, \tag{6.2}$$

$$x_t \leq k_t y_t, \quad t = 1, \ldots, T, \tag{6.3}$$

$$i_0 = 0, x_t, i_t \geq 0, \quad t = 1, \ldots, T, \tag{6.4}$$

$$y_t \in \{0, 1\}, \quad t = 1, \ldots, T. \tag{6.5}$$

The model and its interpretation are the same as the UELSP with the exception of constraint set (6.3), where the production level in period t is limited by the capacity k_t if a production setup occurs in period t (i.e., if $y_t = 1$). Observe that unlike the UELSP, the feasibility of the CELSP depends on the capacity and demand values for a problem instance. In particular, a feasible solution exists for the CELSP if and only if $\sum_{t=1}^{\tau} k_\tau \geq \sum_{t=1}^{\tau} d_\tau$ for every $\tau = 1, \ldots, T$. That is, cumulative capacity must be at least as great as cumulative demand in any period (if shortages are permitted, then we simply require that this condition holds for $\tau = T$). As with the uncapacitated version discussed in the previous chapter, we can create a tighter formulation than the CELSP formulation above. Starting with the tight UELSP$''$ formulation (5.13)–(5.16), if we add the capacity constraints $\sum_{\tau=t}^{T} d_\tau x_{t,\tau} \leq k_t y_t$, for $t = 1, \ldots, T$, to this formulation,[1] then this leads to a tighter formulation of the CELSP (in the sense that the lower bound provided by the LP relaxation of this formulation is at least as high as that provided by the LP relaxation of the above CELSP formulation). As we will see in Chapter 8, the resulting formulation is a spe-

[1]Note that we do not eliminate the upper bounding constraint set (5.15) when adding these capacity constraints to create the tighter formulation. Although these constraints become redundant for the MILP formulation, they are not redundant for the LP relaxation. To see why this is the case, consider a problem instance in which $d_\tau = k_t/2$ for some $\tau \geq t$, and note that a solution with $x_{t,\tau} = 1$ and $y_t = 1/2$ is feasible for the capacity constraint in the LP relaxation, but is not feasible for the constraint $x_{t,\tau} \leq y_t$.

cial case of the discrete facility location problem (FLP). Exercise 6.1 asks the reader to show that the LP relaxation of the above formulation (6.1)–(6.5) admits solutions that the LP relaxation of the "facility location" formulation of the CELSP does not, while the converse is not the case.

Example 6.1 Consider the problem from Example 5.1 with the addition of a production capacity of 70 units in each period. The optimal LP relaxation solution of the CELSP formulation (6.1)–(6.5) has an objective function value of 395.84, while the tighter facility location formulation LP relaxation has an optimal objective function value of 474.2. The optimal MILP solution value equals 489.15. Thus, the facility location formulation, in this case, reduces the integrality gap of the LP relaxation from 19.08% to 3.06%. □

6.3 Relation to the 0-1 Knapsack Problem

Florian, Lenstra, and Rinnooy Kan [50] demonstrated the $\mathcal{NP}$-Hardness of the CELSP through a reduction from the subset sum problem, which, as we discussed in Section 2.2.1, is a special case of the knapsack problem. The subset sum problem considers a set $\mathcal{S}$ of integers $\{s_1, s_2, \ldots, s_n\}$ and asks whether a subset $\mathcal{J} \subseteq \mathcal{S}$ exists such that $\sum_{j \in \mathcal{J}} s_i = S$ for some integer S. We consider a special case of the CELSP such that $f_t = 1$, $h_t = 0$, $k_t = s_t$, and $c_t = (s_t - 1)/s_t$ for all $t = 1, \ldots, T$, where all s_t values are integers. Suppose next that $d_t = 0$ for $t = 1, \ldots, T-1$, and $d_T = S$ for some integer S. Observe that the total production cost in period t equals $1 + [(s_t - 1)/s_t]x_t$ if $x_t > 0$ and equals zero if $x_t = 0$. At $x_t = s_t$, the total cost in period t equals s_t, and when $x_t = 0$ the total cost equals zero, so that at $x_t = 0$ and $x_t = s_t$, the total cost in period t equals x_t for $t = 1, \ldots, T$. For any x_t strictly between 0 and s_t, we have a total cost in the period that exceeds x_t. Thus, for any feasible solution, the total cost is at least as great as $\sum_{t=1}^{T} x_t$.

For this special case of the CELSP, we may ask the question: Does a solution exist with cost at most S? For the special case of the CELSP described, we know that the cost of any feasible solution is at least as great as $\sum_{t=1}^{T} x_t$, and that an optimal solution exists such that $\sum_{t=1}^{T} x_t = S$ (i.e., where total production equals total demand). In addition, for any solution in which production in a period is strictly between 0 and capacity, the total cost of this solution must strictly exceed S (if $0 < x_t < s_t$ in some period t then the cost in this period is strictly greater than x_t; because the cost in each period is at least as great as x_t, this implies that the total cost of such a solution is strictly greater than $\sum_{t=1}^{T} x_t = S$). Thus, a solution with total cost S exists if and only if production in each period t equals either 0 or s_t, as the total cost of such a solution must equal $\sum_{t=1}^{T} x_t = S$. This may occur if and only if a subset of the capacity values $s_1, s_2, \ldots, s_T$ exists with a sum equal to S. This special case of the CELSP is therefore equivalent to the subset sum problem, which implies the $\mathcal{NP}$-Hardness of the CELSP. Note that the optimization version of this special case of the CELSP is equivalent to the 0-1 knapsack problem formulated in Section 2.2.

6.3.1 Fixed-charge network flow interpretation

As with the UELSP, we can represent the CELSP as a fixed-charge network flow problem. In the capacitated case, the production arcs in the network have upper bounds on flows corresponding to the production capacity limits in the associated periods (see Figure 6.1). Because the objective of the CELSP is a concave function, an optimal extreme point solution exists, and we can utilize the network in Figure 6.1 to characterize the structure of extreme point solutions. Recall that extreme point solutions for network flow problems correspond to solutions in which flow balance is achieved and the subnetwork induced by the basic arcs (those with flow strictly between their lower and upper bounds) does not contain any undirected cycles. In contrast to the UELSP, in the CELSP, the production arcs may be nonbasic with positive flow at their *upper bounds*. As a result, extreme point solutions are not unlikely to exist such that production occurs in some period t, i.e., $x_t > 0$, while inventory is held from the prior period, i.e., $i_{t-1} > 0$. That is, extreme point solutions are likely to exist that violate the ZIO property.

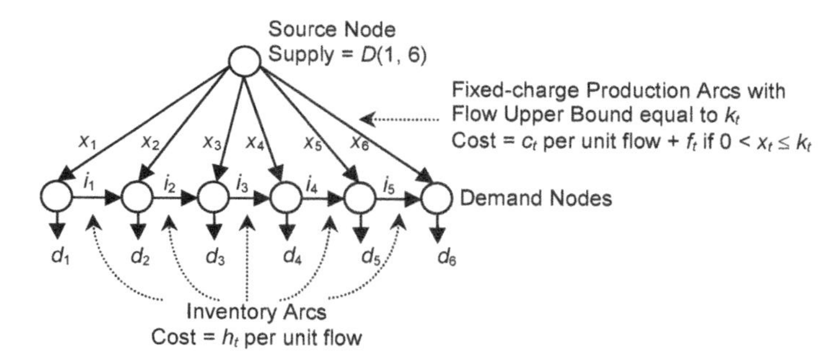

FIGURE 6.1
Fixed-charge network flow representation for a six-period instance of the CELSP.

Clearly, we can characterize any feasible solution to the CELSP as a sequence of RIs, and based on the extreme point properties of network flow problems, we have the following property.

Property 1 *Within a regeneration interval (RI), at most one production period may exist in an extreme point solution with a production level strictly between zero and the period's production capacity.*

To see why Property 1 holds, consider an RI defined by the pair of periods

(s, u), and recall that by the definition of an RI, we have $i_t > 0$ for $t = s, s+1, \ldots, u-2$. Thus, all arcs corresponding to the variables $i_s, i_{s+1}, \ldots, i_{u-2}$ have positive flow and are, therefore, basic arcs. Suppose that two periods, t_1 and t_2, exist within the RI (with $u - 1 \geq t_2 > t_1 \geq s$) such that $0 < x_{t_1} < k_{t_1}$ and $0 < x_{t_2} < k_{t_2}$. Then the basic arcs $x_{t_1}, i_{t_1}, i_{t_1+1}, \ldots, i_{t_2-1}, x_{t_2}$ form a cycle in the corresponding network, and the solution cannot correspond to an extreme point. This implies the following corollary.

Corollary 1 *Within any RI, the production level in each period except at most one must equal either 0 or the period's capacity level.*

Corollary 1 provides an important structural property of optimal solutions for the CELSP, which permits a reduction in the solution space we must search for an optimal solution. This property will be vitally important in developing an efficient solution method for the special case of the CELSP with stationary production capacities, i.e., when $k_t = k$ for all $t = 1, \ldots, T$. Florian and Klein [49] showed that a solution to the CELSP is an extreme point solution if and only if it consists of a sequence of RIs that satisfy Property 1, and also provided the first polynomial time solution approach for the equal-capacity version of the CELSP, which we discuss in Section 6.4.

6.3.2 Dynamic programming approach

Because an optimal solution exists for the CELSP consisting of a sequence of RIs that satisfy Property 1, we can employ the same structural approach to solving the CELSP that we used for the UELSP. That is, using a suitable redefinition of the function $g(t_i, t_{i+1})$, which denotes the minimum cost solution for the RI defined by the pair of periods (t_i, t_{i+1}) (with $t_{i+1} > t_i$), we can apply the recursive equation (5.7) to obtain an optimal solution, $G(1)$, as well as the corresponding shortest path solution approach discussed in Section 5.2.2 and illustrated in Figure 5.2. Unfortunately, in general, it can be very difficult to determine the arc costs of the shortest path graph shown in Figure 5.2 (i.e., the values of $g(t_i, t_{i+1})$ for all (t_i, t_{i+1}) pairs), as they cannot be computed as easily as in the uncapacitated case (i.e., they cannot be computed as stated in the caption of the figure). That is, computing $g(t_i, t_{i+1})$ for specific values of t_i and t_{i+1} may be as difficult as solving the CELSP itself.

Instead of applying the recursive equation (5.7), we consider a different dynamic programming approach, which permits demonstrating that a pseudopolynomial time algorithm exists for the CELSP with integer data (this approach is from Florian et al. [50], who also demonstrate that the CELSP is $\mathcal{NP}$-Hard, although not in the strong sense, as a pseudopolynomial time algorithm exists). To describe this dynamic programming approach, we first define $X_t = \sum_{\tau=1}^{t} x_\tau$ as the cumulative production up to and including period t. Similarly, let $D_t = \sum_{\tau=1}^{t} d_\tau$ denote the cumulative demand up to and including period t, and let $C_t(X)$ denote the minimum cost production plan for periods 1 through t when $X_t = X$; clearly, the optimal solution value for the

CELSP equals $C_T(D_T)$. Beginning with $C_0(0) = 0$ (and setting $C_0(X) = \infty$ for $X \neq 0$), we can recursively compute values of $C_t(X)$ using

$$\begin{aligned} C_t(X) &= \min\{C_{t-1}(X) + h_t(X - D_t); \\ &\quad f_t + h_t(X - D_t) + \min_{x \in \{1,2,\ldots,k_t\}} \{c_t x + C_{t-1}(X - x)\}\}, \end{aligned} \quad (6.6)$$

for a feasible cumulative production quantity X in period t. Note that the feasible cumulative production quantities in period t consist of the set of values $D_t, D_t + 1, \ldots, D_T$. The first term in (6.6) corresponds to the cost if no setup occurs in period t, while the second term computes the minimum cost solution when a setup occurs in period t. For a given t, computing (6.6) requires $\mathcal{O}(k_t D_T)$ time and, therefore, computing an optimal solution for the CELSP using this approach requires $\mathcal{O}((k_1 + k_2 + \ldots + k_T)D_T) = \mathcal{O}(K_T D_T)$ time, where K_T is the cumulative production capacity over all T periods. This approach is, therefore, pseudopolynomial in both K_T and D_T, both elements of problem data.

6.4 The Equal-Capacity Case

While the CELSP is $\mathcal{NP}$-Hard, the practical case in which production capacity is the same in every period (the so-called equal-capacity case, where $k_t = k$ for all $t = 1, \ldots, T$) turns out to be solvable in polynomial time, as originally shown by Florian and Klein [49]. To see this, let us suppose that we can use the shortest path solution approach for the UELSP discussed in Section 5.2.2. Recall that if we knew all of the arc costs, then we could solve the shortest path problem illustrated in Figure 5.2 in $\mathcal{O}(T^2)$ time, where $\mathcal{O}(T^2)$ serves as a bound on the number of arcs in the shortest path network (the actual number of arcs is $T(T+1)/2$). If we can compute the cost of any arc in polynomial time, then clearly we can solve the corresponding problem in polynomial time. Computing the cost of an arc requires determining the minimum cost solution within an RI. In particular, consider the subproblem corresponding to arc (s, u), with $u > s$. To employ the shortest path approach described in Section 5.2.2, arc (s, u) should be labeled with the minimum cost solution to a problem in which inventory is zero in periods $s - 1$ and $u - 1$ and is strictly positive in periods $s, s + 1, \ldots, u - 2$.

To solve the subproblem for the RI (s, u), which corresponds to a $t_{su} = u - s$ period problem, suppose we use the recursive equation (6.6) by reindexing the periods within the RI such that $s = 1$, $T = t_{su}$ (and therefore $u = t_{su} + 1$; this is equivalent to subtracting $s - 1$ from each of the period numbers for all periods from s to $u - 1$ and considering only this t_{su}-period problem). The recursive approach implied by equation (6.6) requires evaluating cumulative production levels over time. Because of Corollary 1, and because we wish to determine the minimum cost solution such that (s, u) corresponds to an RI, we have a limited number of cumulative production levels that we need to consider. That is, we know that for all periods with positive production, at most one of these does not have a production level of k. And, because $i_{s-1} = i_{u-1} = 0$, we know that the total production within the RI must equal the total demand within the RI. In particular, for the t_{su}-period problem, we must have $X_{t_{su}} = D_{t_{su}}$. Given that at most one period can have a production level that does not equal k, we also know that the production level in the corresponding period must equal ϵ, where

$$\epsilon = D_{t_{su}} - k \left\lfloor \frac{D_{t_{su}}}{k} \right\rfloor, \tag{6.7}$$

i.e., the remainder when $D_{t_{su}}$ is divided by k. This implies that for X_1, we need to consider the cumulative production quantities 0, ϵ, and k only (clearly we do not need to consider $X_1 = 0$ if $D_1 > 0$, nor do we need to consider $X_1 = \epsilon$ if $D_1 > \epsilon$; thus, when stating the number of cumulative production quantities we need to consider in a period, we are taking a worst-case approach). For X_2, we need to consider cumulative production quantities 0, ϵ, k, $k + \epsilon$, and $2k$. As a result, we can rewrite our recursive equation as

$$\begin{aligned} C_t(X) &= \min\{C_{t-1}(X) + h_t(X - D_t); \\ &\quad f_t + h_t(X - D_t) + \min_{x \in \{0, \epsilon, k\}} \{c_t x + C_{t-1}(X - x)\}\}, \end{aligned} \tag{6.8}$$

where (in the worst-case) we must consider all $X \in \{0, \epsilon, k, k + \epsilon, 2k, 2k + \epsilon, \ldots, nk, nk + \epsilon\}$, where $n = \lfloor D_{t_{su}}/k \rfloor$. Clearly $n \leq T$, which implies that, in the worst case, we need to evaluate $\mathcal{O}(T)$ values of X, and we can thus compute all required values of $C_t(X)$ for any t in $\mathcal{O}(T)$ time. Because there are $\mathcal{O}(T)$ values of t for any RI subproblem, computing $C_t(X)$ for all values of t for an RI subproblem (and, therefore, determining $C_{t_{su}}(D_{t_{su}})$, an optimal RI solution) requires $\mathcal{O}(T^2)$ time using the recursive equation (6.8). Because we must determine an optimal RI solution for $\mathcal{O}(T^2)$ RIs, where each such solution corresponds to the cost of an arc in the shortest path problem, the overall worst-case complexity of this solution approach is $\mathcal{O}(T^4)$. Note that van Hoesel and Wagelmans [111] demonstrated how to reduce this worst-case complexity to $\mathcal{O}(T^3)$ by exploiting structural relationships between different RI subproblem solutions.

Example 6.2 Returning to the problem from Example 6.1, the resulting shortest path network is shown in Figure 6.2. Above each arc (s, u), the figure shows the minimum cost associated with the RI (s, u), while the bold arcs highlight the shortest path in the network, which has a cost equal to 489.15. The optimal path consists of RIs $(1, 3)$ and $(3, 6)$. The optimal RI $(1, 3)$ solution produces 63 units in period 1 and 70 units in period 2, while the optimal RI $(3, 6)$ solution produces 59 units in period 3, 70 units in period 4, and 0 units in period 5. Exercise 6.5 asks the reader to compute the arc costs shown in Figure 6.2. Note that no arc exists from node 2 to higher numbered nodes. This results because $d_2 > k_2$ and, therefore, no feasible RI exists beginning at period 2 (the minimum cost for each such RI thus equals ∞). □

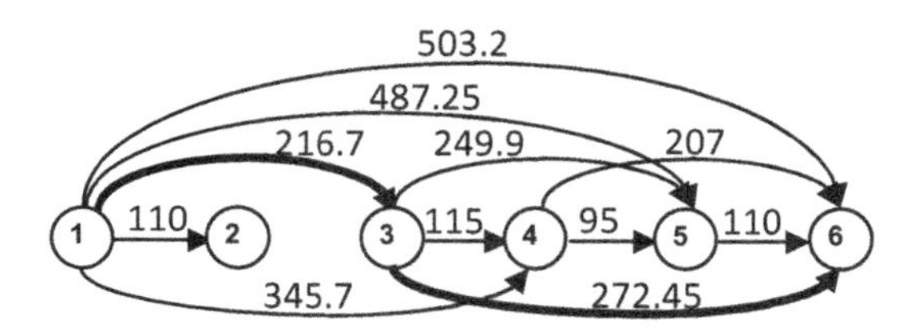

FIGURE 6.2
Shortest path network resulting from dynamic programming solution of the equal-capacity CELSP for the problem from Example 6.1.

6.5 FPTAS for Capacitated Lot Sizing

Section 2.5 introduced the idea of a fully polynomial time approximation scheme (FPTAS) for the $\mathcal{NP}$-Complete knapsack problem. Recall that an FPTAS provides a solution that is within $\epsilon \times 100\%$ of the optimal solution value in polynomial time (as a function of the problem size and $1/\epsilon$ for a fixed $\epsilon > 0$). This section describes the FPTAS for the CELSP provided by van Hoesel and Wagelmans [112]. For ease of exposition, we present the analysis for the CELSP as formulated in (6.1) - (6.5), i.e., with linear holding costs, no shortages allowed, and fixed plus linear production cost functions (van Hoesel and Wagelmans [112] considered a more general case involving concave production cost functions, and concave holding and backordering cost functions).

6.5.1 Structure of the dynamic programming approach

Developing an FPTAS for the CELSP requires a dynamic programming problem formulation that is fundamentally different from the ones discussed in Sections 6.3.2 and 6.4. Instead of the state of the dynamic program being defined explicitly by a set of decision variables, van Hoesel and Wagelmans [112] use a recursion in which the state is effectively the maximum budget, or cost, that may be incurred by the end of a time period. And, instead of a value function that is defined in terms of cost, they use a value function that represents the maximum inventory level at the end of a time period, given that the budget is not exceeded. Using such an approach requires an assumption that all problem parameters take integer values, and that we have some known integer upper bound on the problem's optimal solution.

Suppose we have an upper bound B on the problem's optimal solution value, and let $F_t(b)$ denote the maximum value of inventory at the end of period t (i_t) that can be achieved while spending no more than b ($\leq B$) of the budget by the end of period t. Using this construction, then the optimal solution value for the problem up to the end of period t can be obtained by determining the smallest value of b such that $F_t(b) \geq 0$, i.e., the smallest amount of budget spent while ensuring that all demands are met. Observe that for a given b, if no feasible solution exists with cost less than or equal to b, then we set $F_t(b) = -\infty$, which will ensure that such combinations of t and b can never be part of an optimal solution.

In order to provide a dynamic programming recursive relationship, we would like to be able to decompose the problem into a set of smaller subproblems. Suppose that in period t, we know $F_{t-1}(a)$ for some $a \leq b$, and we would like to determine $F_t(b)$. The next question is whether we can use the value of $F_{t-1}(a)$ and the production and holding costs in period t only to determine $F_t(b)$. Given that we have allocated a units of the budget to periods prior to period t, this leaves $b - a$ units of budget for the costs incurred in period t.

Certainly we can enumerate the possible production levels in period t from among the set of values $x_t = \{0, 1, \ldots, k_t\}$, and determine the cost in period t for each such solution, assuming that the inventory at the end of period $t-1$ equals $F_{t-1}(a)$. (Note that we will assume that the cost of any infeasible solution is infinite, i.e., any solution with negative inventory at the end of period t.) If we compute the cost associated with period t for each possible production level, then two situations can arise. In the first situation, we may find that no solution exists such that the cost in period t is less than or equal to the budget available to period t, which equals $b - a$. The second situation considers the case in which a solution does exist with cost in period t less than or equal to $b - a$.

Suppose that the latter situation occurs, i.e., we are able to find a production level in period t such that the production and holding costs in period t are less than or equal to $b - a$. Van Hoesel and Wagelmans [112] show that it is sufficient to consider only solutions such that $i_{t-1} = F_{t-1}(a)$ when computing $F_t(b)$ using the following arguments. Suppose a solution exists with period-t cost less than or equal to $b - a$ and with $i_{t-1} = i'_{t-1} < F_{t-1}(a)$. Let x'_t be the maximum possible production in period t when $i_{t-1} = i'_{t-1}$, such that the production and holding costs in period t are less than or equal to $b - a$. Because a solution exists with $i_{t-1} = F_{t-1}(a)$ and with production and holding cost in period t less than or equal to $b - a$, we know that $x'_t > 0$.[2]

[2]To show that $x'_t > 0$, suppose we take the solution with $i_{t-1} = F_{t-1}(a)$ and assume that the production level in period t equals x_t. Then if $F_{t-1}(a) + x_t > d_t$, and we reduce i_{t-1}, which also deccreases i_t, we free up some of the budget $b - a$ to allow for more production in period t while staying within this budget. If, on the other hand, $F_{t-1}(a) + x_t = d_t$, then the solution with $i'_{t-1} < F_{t-1}(a)$ must have had $x'_t > 0$ in order to cover demand in period

Now let $x''_t = x'_t - \min\{F_{t-1}(a) - i'_{t-1}; x'_t\}$ and observe that $0 \leq x''_t < x'_t$. Consider the solution with $i_{t-1} = F_{t-1}(a)$ and $x_t = x''_t$. The holding cost in period t associated with this solution equals the holding cost associated with the solution such that $x_t = x'_t$ and $i_{t-1} = i'_{t-1}$, and the associated production cost in period t is lower for the solution with $x_t = x''_t$. Thus, the solution with $x_t = x''_t$ and $i_{t-1} = F_{t-1}(a)$ also satisfies the budget constraint in period t of $b - a$. This implies that for any solution that satisfies the period-t budget constraint with $i_{t-1} < F_{t-1}(a)$, we can construct a solution that satisfies the budget constraint with $i_{t-1} = F_{t-1}(a)$, i.e., if a solution exists that satisfies the budget constraint in period t, then we can limit our attention to solutions such that $i_{t-1} = F_{t-1}(a)$ in determining $F_t(b)$.

Now suppose that no solution exists with $x_t \in \{0, 1, \ldots, k_t\}$, $i_{t-1} = F_{t-1}(a)$, and with period-t production and holding costs less than or equal to $b - a$. Van Hoesel and Wagelmans show that when this is the case, only solutions such that $x_t = 0$ and $i_{t-1} \in \{d_t, d_t + 1, \ldots, F_{t-1}(a) - 1\}$ need to be considered in order to determine $F_t(b)$ for $b > a$. To show this, consider a solution such that $i_{t-1} = i'_{t-1} < F_{t-1}(a)$ and $x'_t > 0$ that satisfies the budget constraint and gives the maximum possible inventory at the end of period t. Note that if $i'_{t-1} + x'_t \geq F_{t-1}(a)$, and this solution is feasible for the budget $b - a$, then a solution with $i''_{t-1} = F_{t-1}(a)$ and $x''_t = x'_t - F_{t-1}(a) + i'_{t-1}$ is also feasible for the budget constraint (because $x''_t < x'_t$, and the inventory at the end of period t is the same in both solutions). This contradicts the claim that no solution exists with $i_{t-1} = F_{t-1}(a)$ that satisfies the period-t budget constraint of $b - a$; thus, we must have $i'_{t-1} + x'_t < F_{t-1}(a)$. This further implies that we cannot have $F_{t-1}(a) \leq d_t$, because $i'_{t-1} + x'_t < F_{t-1}(a)$, and these two inequalities taken together give $i'_{t-1} + x'_t < d_t$, which implies an inability to satisfy demand in period t (which corresponds to an infeasible solution and, hence, infinite cost).

We therefore have that $i'_{t-1} + x'_t$ must be in the interval $[d_t, F_{t-1}(a))$. It should be clear, however, that any solution in which $i_{t-1} \in [d_t, F_{t-1}(a))$ has a cost of at most a in periods one through $t - 1$ (reducing the end-of-period $t - 1$ inventory simply reduces the cost associated with the first $t - 1$ periods). Consider, therefore, a solution with $i''_{t-1} = i'_{t-1} + x'_t$ and $x''_t = 0$. Clearly, because $i''_{t-1} = i'_{t-1} + x'_t < F_{t-1}(a)$, the cost associated with such a solution is no more than a in the first $t - 1$ periods. Moreover, the cost in period t associated with this solution cannot be greater than that of the solution with $i_{t-1} = i'_{t-1}$ and $x_t = x'_t > 0$ (which incurs positive production cost in period t), and both solutions have the same (maximal) ending inventory in period t. As a result, the solution x''_t and i''_{t-1} must satisfy the budget constraint with maximal inventory at the end of period t. This implies that we only need to consider solutions with $x_t = 0$ to determine $F_t(b)$ when no solution satisfies the budget with $i_{t-1} = F_{t-1}(a)$.

---

t. In either case, the maximum production level such that the cost in period t is less than or equal to $b - a$ must be strictly positive.

The preceding two paragraphs lead to the conclusion that for a given value of budget b, we can determine $F_t(b)$ by considering two types of solutions: (i) those that consume a budget of $a \leq b$ up to period $t-1$ with $i_{t-1} = F_{t-1}(a)$; and (ii) those with zero production in period t and with a nonnegative inventory level at the end of period t that is less than $F_{t-1}(a) - d_t$ for $a \leq b$. We formalize the recursive relationship as follows. Let $p_t(x_t)$ denote the production cost in period t, where $p_t(x_t) = \mathbf{1}_{\{x_t>0\}}(f_t + c_t x_t)$ (the value of $\mathbf{1}_{\{x_t>0\}}$ equals one if $x_t > 0$, and zero otherwise). Define $I_t(x_t, a) = F_{t-1}(a) + x_t - d_t$ and let $\mathcal{F}_t(b|a)$ denote, for fixed a and b with $b \geq a$, the maximum inventory at the end of period t while using at most a budget of a in periods one through $t-1$ and a budget of at most b in periods one through t. We compute $\mathcal{F}_t(b|a)$ using

$$\mathcal{F}_t(b|a) = \max \left\{ \begin{array}{c} \max\limits_{0 \leq x_t \leq k_t} \{I_t(x_t, a) | p_t(x_t) + h_t I_t(x_t, a) \leq b - a\}, \\ \max\limits_{0 \leq i_t < F_{t-1}(a) - d_t} \{i_t | h_t i_t \leq b - a\} \end{array} \right\}. \quad (6.9)$$

Then we can find $F_t(b)$ using

$$F_t(b) = \max_{0 \leq a \leq b} \{\mathcal{F}_t(b|a)\}. \quad (6.10)$$

The top expression in (6.9) can be determined for fixed values of a and b in $\mathcal{O}(\log k_t)$ time using a binary search (because production and holding costs are monotonically increasing), and the bottom expression requires $\mathcal{O}(\log \sum_{\tau=1}^{t-1} k_\tau)$ operations (using binary search and noting that $F_{t-1}(a) - d_t - 1$ cannot exceed $\sum_{\tau=1}^{t-1} k_\tau$). Because the top expression must be evaluated $\mathcal{O}(B^2)$ times, and the second expression can be determined by performing the binary search at most B times for each value of $t = 1, \ldots, T$, the complexity of the overall dynamic programming solution is $\mathcal{O}(B^2 \sum_{t=1}^{T} \log k_t + B \sum_{t=1}^{T} \log \sum_{\tau=1}^{t-1} k_\tau)$.

6.5.2 Approximation of the dynamic program

The dynamic program discussed in the previous subsection has a running time that is pseudopolynomial in the upper bound on the objective function value B. As with the FPTAS discussed for the knapsack problem in Section 2.5, we will effectively scale the objective (or budget) by some integer value $D \geq 1$ (clearly it is not an approximation algorithm when $D = 1$, and we will therefore focus on values of D greater than 1). In other words, instead of evaluating the dynamic program for all values $b = 0, 1, 2, \ldots, B$, suppose we evaluate it for values of $b = D, 2D, \ldots, \bar{D}$ for some integer $\bar{D}$. We then let $F_t^D(b)$ correspond to the maximum inventory level at the end of period t under a budget of b, which must be a multiple of D. Among all solutions with $b \in \{D, 2D, \ldots, \bar{D}\}$, we seek the smallest b such that $F_T^D(b) \geq 0$.

In order to use this approach, we need to know an appropriate value for $\bar{D}$, and we need to ensure that we are able to find a feasible solution for the underlying CELSP. To determine an appropriate value of $\bar{D}$, suppose we know that an optimal solution allocates a budget of δ_t to period t for $t = 1, \ldots, T$, i.e., the optimal solution value, denoted by z^*, equals $\sum_{t=1}^{T} \delta_t$. If we were to allocate a budget of $(\lfloor \delta_t/D \rfloor + 1)D$ to each period t, then the optimal solution would satisfy these budgets. Moreover, the allocation of these budgets would ensure that we evaluate $F_T^D(b)$ at $b = \sum_{t=1}^{T}(\lfloor \delta_t/D \rfloor + 1)D$ (a multiple of D), and that a feasible solution exists for this value of budget such that $F_T^D(b) \geq 0$. Although we do not know z^* or the corresponding values of δ_t, for $t = 1, \ldots, T$, we do know that $z^* \leq B$. We also know that $\sum_{t=1}^{T}(\lfloor \delta_t/D \rfloor + 1)D \leq (\lfloor \sum_{t=1}^{T} \delta_t/D \rfloor + T)D = (\lfloor z^*/D \rfloor + T)D \leq (\lfloor B/D \rfloor + T)D$. This implies that if we set $\bar{D} = (\lfloor B/D \rfloor + T)D$, we ensure that a feasible solution exists among $b \in \{D, 2D, \ldots, \bar{D}\}$ with $F_T^D(b) \geq 0$. It also ensures that we find a solution with objective value bounded from above by $(\lfloor z^*/D \rfloor + T)D \leq z^* + DT$. The complexity of this approximation algorithm is also bounded by $\mathcal{O}((B/D + T)^2 \sum_{t=1}^{T} \log k_t + (B/D + T) \sum_{t=1}^{T} \log \sum_{\tau=1}^{t-1} k_\tau)$.

In order to arrive at an FPTAS for the CELSP, we need to ensure that we are able to determine an upper bound on the objective function value in polynomial time whose bound is stated in terms of the optimal solution value, z^*. To do this, suppose we can find the smallest number L such that a feasible solution is guaranteed to exist with a production cost in any period less than or equal to L and a holding cost in any period less than or equal to L as well. If we are able to find such a number, then we know that the objective function value is no more than $2TL$. And, because the optimal solution value is z^*, we know that a feasible solution exists with a maximum production cost of z^* in any period and a maximum holding cost of z^* in any period as well. Therefore, we know that $L \leq z^*$, which implies $2TL \leq 2Tz^*$. As a result, if we can find the value of L in polynomial time, we can then provide a bound on the objective function that is no more than $2Tz^*$.

Given an arbitrary value l, let $x_t^u = \max\{x_t \leq k_t | p_t(x_t) \leq l\}$ and $i_t^u = \max\{i_t | h_t i_t \leq l\}$ denote, respectively, the maximum production and inventory levels in period t such that the production and inventory costs in period t do not exceed l. These production and inventory bounds can easily be determined in polynomial time for any l. We can then determine, via dynamic programming, whether, for a given l, a feasible solution exists satisfying all of these bounds on production and inventory levels. And, given an upper bound on L, we can then determine the value of L via binary search. (An initial upper bound on L is given by $\max_{\{t=1,\ldots,T\}}\{p_t(k_t), h_t(\sum_{\tau=t+1}^{T} d_\tau)\}$.)

We can therefore compute an upper bound B in polynomial time, such that $B \leq 2Tz^*$. Next, suppose we set our scaling factor, D, such that $D = \max\left\{\lfloor \epsilon B/2T^2 \rfloor, 1\right\}$ for some desired optimality tolerance $\epsilon > 0$, and apply the approximate (scaled) dynamic programming approach. Recall that the complexity of the dynamic programming approximation algorithm is $\mathcal{O}((B/D + T)^2 \sum_{t=1}^{T} \log k_t + (B/D + T) \sum_{t=1}^{T} \log \sum_{\tau=1}^{t-1} k_\tau)$, which is polynomial in T if $B/D \leq T$. If $B/D > T$, then either $D = 1$ or $D = \lfloor \epsilon B/2T^2 \rfloor$. If $D = 1$, then this implies $\epsilon B/2T^2 \leq 1 = D$, i.e., $B/D \leq 2T^2/\epsilon$. If $D = \lfloor \epsilon B/2T^2 \rfloor$, then $2D > \epsilon B/2T^2$, i.e., $B/D < 4T^2/\epsilon$. In both of these cases, the value of B/D is bounded by $\mathcal{O}(T^2/\epsilon)$. The important result is then that the complexity of the dynamic programming approximation algorithm becomes

$$\mathcal{O}\left(\frac{T^4}{\epsilon^2}\sum_{t=1}^{T}\log k_t + \frac{T^2}{\epsilon}\sum_{t=1}^{T}\log\sum_{\tau=1}^{t-1}k_\tau\right), \tag{6.11}$$

which is polynomial in T and $1/\epsilon$. If $D = \lfloor \epsilon B/2T^2 \rfloor$, the bound on the objective function is $z^* + DT \leq z^* + \epsilon B/2T$; because $B \leq 2Tz^*$, this implies that the objective function is bounded by $z^*(1+\epsilon)$. (If $D = 1$, then we guarantee finding z^*.)

6.6 Valid Inequalities for the CELSP

In this section, we would like to gain some intuition on the structure and use of valid inequalities for lot sizing problems. Valid inequalities are often added to the LP relaxation of an MILP in order to eliminate solutions that do not satisfy the integrality requirements, and improve the bound provided by the LP relaxation value. We begin by considering a simple integer rounding technique. First note that feasibility (in the absence of backordering) for the problem requires cumulative production levels that are at least as great as cumulative demand levels in each period, i.e.,

$$\sum_{\tau=1}^{t} x_\tau \geq \sum_{\tau=1}^{t} d_\tau, \qquad t = 1, \ldots, T. \tag{6.12}$$

Given that we also require $x_t \leq k_t y_t$ for $t = 1, \ldots, T$, this implies the validity of the following inequalities:

$$\sum_{\tau=1}^{t} k_\tau y_\tau \geq \sum_{\tau=1}^{t} d_\tau, \qquad t = 1, \ldots, T. \tag{6.13}$$

We next define $\overline{k}_t = \max_{\tau=1,\ldots,t} k_\tau$, and note that $\overline{k}_t \sum_{\tau=1}^{t} y_\tau \geq \sum_{\tau=1}^{t} k_\tau y_\tau$. Taken together with (6.13), and noting that $\sum_{\tau=1}^{t} y_\tau$ must be integer, we have

$$\sum_{\tau=1}^{t} y_\tau \geq \left\lceil \frac{\sum_{\tau=1}^{t} d_\tau}{\overline{k}_t} \right\rceil, \qquad t = 1, \ldots, T. \tag{6.14}$$

We can generalize this approach as follows. Suppose that for a given t, we reindex periods from 1 to t in nonincreasing order of production capacity levels such that $k_{[1]}^t \geq k_{[2]}^t \geq \ldots k_{[t]}^t$, where $k_{[i]}^t$ denotes the capacity of the period with the ith highest capacity level between periods 1 and t inclusive. After this reindexing, let m denote the smallest value of i from 1 to t such that $\sum_{\tau=1}^{i} k_{[i]}^{t} \geq \sum_{\tau=1}^{t} d_\tau$. Then we can replace the (entire) right-hand side of (6.14) with the resulting value of m. The strength and usefulness of these simple rounding-based inequalities depends to a large degree on the amount of variation in the production capacity levels and the relationship between these capacity levels and demand values.

Example 6.3 Example 6.1 noted that the integrality gap of the tight facility location formulation for the problem instance equals 3.06%, with an LP relaxation value of 474.2 (the optimal CELSP solution value for this instance equals 489.15). If we add the valid inequality (6.14) with $t = 5$, i.e., $\sum_{\tau=1}^{5} y_\tau \geq \left\lceil \sum_{\tau=1}^{5} d_\tau/70 \right\rceil = 4$, then the LP relaxation solution value increases to 485.81, giving an integrality gap of 0.68%. Noting, however, that $\left\lceil \sum_{\tau=1}^{4} d_\tau/70 \right\rceil = 4$, the rounding inequality with $t = 4$, which is given by $\sum_{\tau=1}^{4} y_\tau \geq 4$, is stronger than that with $t = 5$; when this inequality is used, the LP relaxation solution value equals 489.15, and the integrality gap is closed to 0. □

6.6.1 (S, l) inequalities

We next consider a set of valid inequalities for the UELSP that are used in describing the convex hull of feasible solutions for the UELSP. To motivate these inequalities, consider inequality (6.12) for some value of $t < T$. This inequality can be strengthened by considering whether or not production in periods 1 through t serves demand in period $t+1$. If $y_{t+1} = 0$, then this must be the case. Otherwise, if $y_{t+1} = 1$, the demand in period $t+1$ may be served by the setup in period $t+1$. Letting $d_{1,t} = \sum_{\tau=1}^{t} d_\tau$ (with a slight abuse of notation), this leads to the inequality

$$\sum_{\tau=1}^{t} x_\tau \geq d_{1,t} + d_{t+1}(1 - y_{t+1}), \tag{6.15}$$

or

$$\sum_{\tau=1}^{t} x_\tau + d_{t+1} y_{t+1} \geq d_{1,t+1}. \tag{6.16}$$

Next consider accounting for period $t+2$. Clearly it is possible that a setup in either period $t+1$ or $t+2$ may satisfy the demand in period $t+2$. Consider the inequality

$$\sum_{\tau=1}^{t} x_\tau \geq d_{1,t} + d_{t+1}(1 - y_{t+1}) + d_{t+2}(1 - y_{t+1} - y_{t+2}). \tag{6.17}$$

If $y_{t+1} = 1$ and $y_{t+2} = 0$, then this inequality reduces to (6.12). If $y_{t+1} = 0$ and $y_{t+2} = 1$, then we require production in periods 1 through t to cover demand in periods 1 through $t + 1$, as desired. If $y_{t+1} = y_{t+2} = 0$ then we require production in periods 1 through t to cover demand in periods 1 through $t+2$, as desired. If $y_{t+1} = y_{t+2} = 1$, then the inequality is dominated by (6.12). In each of these cases (6.17) is valid and can be written as

$$\sum_{\tau=1}^{t} x_\tau + \sum_{\tau=t+1}^{t+2} d_{\tau,t+2} y_\tau \geq d_{1,t+2}. \tag{6.18}$$

More generally, suppose we consider periods 1 through l for some $l \leq T$ and let $L = \{1, 2, \ldots, l\}$. Next define the set $S = \{1, \ldots, t\}$ for some $t < l$. Then we can write a generalized version of (6.18) as

$$\sum_{\tau \in S} x_\tau + \sum_{\tau \in L \setminus S} d_{\tau,l} y_\tau \geq d_{1,l}. \tag{6.19}$$

Inequality (6.19) gives the form of the so-called (S, l) inequalities provided by Barany, Van Roy, and Wolsey [12]. As they show, the set S can be defined as *any* subset of L, and (6.19) remains valid for the UELSP. They also show that the (S, l) inequalities serve as facets of the convex hull of mixed integer solutions for the UELSP. Moreover, they demonstrate that the set of all (S, l) inequalities, along with simple inequalities from the LP relaxation formulation of the UELSP, define the entire convex hull. Because these inequalities are valid for the UELSP, they are therefore also valid for the CELSP (clearly the UELSP is a relaxation of the CELSP).

We can further generalize the (S, l) inequalities by permitting the set L to begin at a period number greater than one. That is, suppose the set L contains the periods $\{u, u+1, \ldots, l\}$ for some $u \geq 1$. As before, the set S may consist of any subset of L. Observe that the demand in periods u through l may be met either by production in these periods, or inventory at the end of period $u - 1$. After our redefinition of the set L, we can write the more general (S, l) inequalities as

$$I_{u-1} + \sum_{\tau \in S} x_\tau + \sum_{\tau \in L \setminus S} d_{\tau,l} y_\tau \geq d_{u,l}. \tag{6.20}$$

Note that for any period τ between u and l, the terms x_τ and $d_{\tau,l} y_\tau$ are essentially substitutable for one another in the (S, l) inequalities. That is, if we move some period τ from the set S to the set $L \setminus S$, we replace the term x_τ with the term $d_{\tau,l} y_\tau$, and vice versa.

6.6.2 Facets for the equal-capacity CELSP

The following valid inequalities, which look similar to the (S, l) inequalities and use our previous definitions of the sets L and S, are implied by the LP relaxation of the equal-capacity version of the CELSP with capacity k in each period:

$$I_{u-1} + \sum_{\tau \in S} x_\tau + k \sum_{\tau \in L \setminus S} y_\tau \geq d_{u,l}. \tag{6.21}$$

(Observe that for the CELSP with unequal capacities, it is valid to substitute $\overline{k}_{L \setminus S} = \max_{\tau \in L \setminus S} \{k_\tau\}$ for k in the above inequality.) We next illustrate how to use (6.21) to derive a set of strong valid inequalities that are not implied by the LP relaxation and, in some cases, serve as facets of the convex hull of mixed integer solutions for the equal-capacity CELSP. We first rearrange (6.21) as follows:

$$I_{u-1} + \sum_{\tau \in S} x_\tau \geq d_{u,l} - k \sum_{\tau \in L \setminus S} y_\tau. \tag{6.22}$$

Define r as the remainder when dividing $d_{u,l}$ by the capacity k using modulo arithmetic, i.e., $r = d_{u,l}(\text{mod} k) = d_{u,l} - k\lfloor d_{u,l}/k \rfloor$. This implies that we can replace $d_{u,l}$ with $r + k\lfloor d_{u,l}/k \rfloor$, and rewrite the right-hand side of (6.22) as

$$r + k \left(\lfloor d_{u,l}/k \rfloor - \sum_{\tau \in L \setminus S} y_\tau \right). \tag{6.23}$$

The term in parentheses in (6.23) must take an integer value (although not necessarily nonnegative). Let p denote this term in parentheses, i.e., $p = \lfloor d_{u,l}/k \rfloor - \sum_{\tau \in L \setminus S} y_\tau$, so that the right-hand side of (6.22) may be written as simply $r + kp$. Next let $X = I_{u-1} + \sum_{\tau \in S} x_\tau$, which is the left-hand side of (6.22). In this case, we can write (6.22) as

$$X \geq r + kp, \tag{6.24}$$

where X is a nonnegative continuous variable, r and k are positive scalars, and p is an integer variable. Figure 6.3 illustrates this inequality graphically in terms of the aggregated variables X and p, where the shaded region corresponds to values of (p, X) that are feasible for the inequality.

The shaded region in the figure does not account for the integrality requirement for the variable p, while the bold arrows in the figure show the solutions that satisfy the integrality requirement for p. Clearly if we rotate the inequality clockwise about the point $(0, r)$ until it hits the point $(-1, 0)$, we eliminate part of the continuous feasible region without eliminating any of the feasible solutions that satisfy the integrality requirement for p (see Figure 6.4; note that for exceptional cases in which $r = 0$, i.e., $d_{u,l}$ is a multiple of k, no such rotation is possible, and the inequality derived below is implied by the LP relaxation). The resulting inequality is given by

$$X \geq r + rp, \tag{6.25}$$

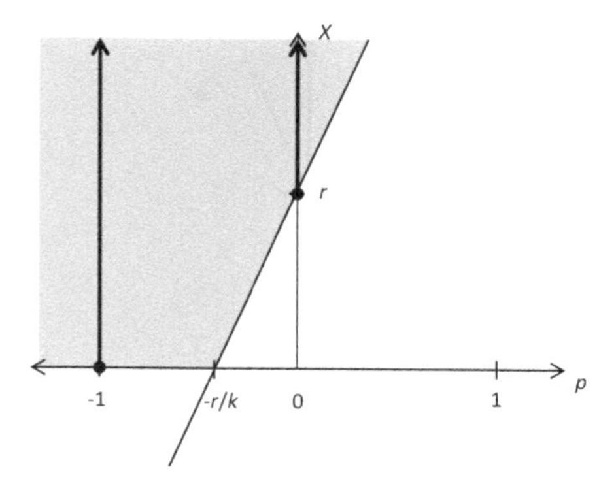

FIGURE 6.3
Illustration of inequalities implied by the LP relaxation formulation in aggregate variable space.

where $r(1+p) = r(\lceil d_{u,l}/k \rceil - \sum_{\tau \in L \setminus S} y_\tau)$. This inequality in expanded form can be written as

$$I_{u-1} + \sum_{\tau \in S} x_\tau \geq r \left(\left\lceil \frac{d_{u,l}}{k} \right\rceil - \sum_{\tau \in L \setminus S} y_\tau \right). \tag{6.26}$$

Next observe that, just as with the (S, l) inequalities, we can effectively substitute any x_τ on the left-hand side of the inequality with $d_{\tau,l} y_\tau$ (which is equivalent to reducing the right-hand side by $d_{\tau,l}$ when $y_\tau = 1$). Letting $D \subseteq S$ denote the set of periods for which this substitution is made, the resulting inequality is given by

$$I_{u-1} + \sum_{\tau \in S \setminus D} x_\tau + \sum_{\tau \in D} d_{\tau,l} y_\tau \geq r \left(\left\lceil \frac{d_{u,l}}{k} \right\rceil - \sum_{\tau \in L \setminus S} y_\tau \right). \tag{6.27}$$

Leung, Magnanti, and Vachani [75] first provided inequality (6.27) under a specially structured definition of the sets $L \setminus S$ and D. Defining these sets requires first considering the set of periods for which a y_τ variable appears in the inequality (as opposed to an x_τ variable). Call this set $\mathcal{Y}$. Next, we need to identify a critical period index w as the smallest period index in L such that $\lceil d_{u,w}/k \rceil = \lceil d_{u,l}/k \rceil$. Then, for any $\tau \in \mathcal{Y}$, if $\tau \leq w$, we insert τ in $L \setminus S$; if $\tau \geq w+1$, we insert τ in D. Under these set definitions and an additional set of requirements on a problem instance's parameter values and the cardinality of the sets S, D, and $L \setminus S$, Leung et al. [75] show that (6.27) provides a facet of the convex hull of solutions for the equal-capacity CELSP. We note that Pochet [93] provided a class of inequalities for the CELSP at around the same time as Leung et al. [75], and demonstrated conditions under which these

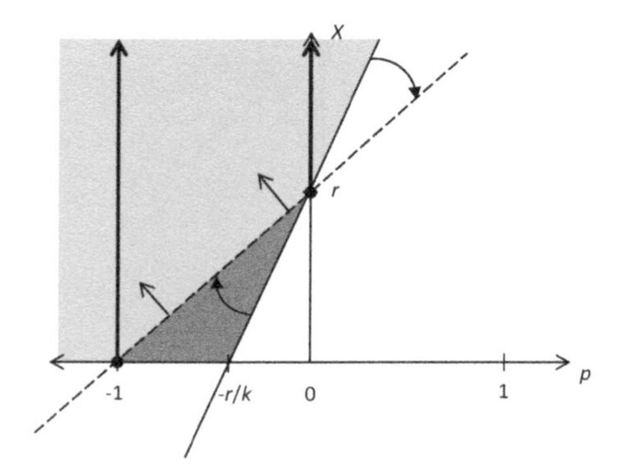

FIGURE 6.4
Rotation of implied inequality cuts off part of the continuous feasible region that does not satisfy the integrality requirement for p.

inequalities serve as facets of the convex hull of solutions for the equal-capacity CELSP. Although these inequalities (in the equal-capacity case) appear quite similar to (6.27), Leung et al. [75] show that they are neither subsumed by (6.27), nor do they dominate (6.27). We will discuss the inequalities provided by Pochet [93] in the next section.

Finally, it is interesting to note that when $S = \emptyset$ and $u = 1$, (6.27) reduces to our simple rounding inequalities (6.14) with $l = t$. Moreover, when $k = \infty$, $r = d_{u,l}$, and (6.27) reduces to the (S, l) inequality (6.20) (noting that it is valid to replace the coefficient of y_t for any $t \in L \backslash S$ with $\min\{r, d_{t,l}\}$, which equals $d_{t,l}$ in the uncapacitated case).

6.6.3 Generalized flow-cover inequalities

A class of inequalities for the CELSP (without equal capacities) follows from so-called *flow-cover* inequalities (see Nemhauser and Wolsey [87]). For this set of inequalities, we again consider a set of periods $L = \{u, u+1, \ldots, l\}$ where $u \geq 1$ and $l \leq T$, as well as a subset of these periods $S \subseteq L$. The CELSP formulation implies

$$\sum_{t \in S} x_t \leq \sum_{t \in L} x_t \leq d_{u,l} + i_l, \tag{6.28}$$

as well as $x_t \leq k_t y_t$ for all $t = 1, \ldots, T$. Analogous to our discussion of the knapsack problem in Chapter 2, we call a set S a *cover* for the demand from period u through l if $\sum_{t \in S} k_t > d_{u,l}$. For such a cover set, let λ equal the *excess capacity* of the cover, i.e., $\lambda = \sum_{t \in S} k_t - d_{u,l} > 0$. Clearly such a set S has the capacity to serve all of the demand $d_{u,l}$; however, if we set $y_\tau = 0$ for some $\tau \in S$, this capacity will be reduced. In particular, if $y_\tau = 0$, then the set $S \backslash \{\tau\}$ is capable of meeting $d_{u,l} - \max\{k_\tau - \lambda; 0\} = d_{u,l} - (k_\tau - \lambda)^+$ units of demand in periods u through l. If we set $y_t = 0$ for all $t \in S' \subset S$, then the set $S \backslash S'$ is capable of meeting $d_{u,l} - (\sum_{t \in S'} k_t - \lambda)^+$ units of demand in periods u through l. Noting that $(\sum_{t \in S'} k_t - \lambda)^+ \geq \sum_{t \in S'}(k_t - \lambda)^+$, then the quantity $d_{u,l} - \sum_{t \in S'}(k_t - \lambda)^+$ serves as an upper bound on the amount of demand in periods u through l that may be met with production in periods within the set $S \backslash S'$. As a result, the following inequalities are valid for the CELSP, where the set S' is determined by those $t \in S$ such that y_t is set to zero (note that the x_t variable associated with any $t \in S$ such that $y_t = 0$ will be forced to zero as a result of the constraint $x_t \leq k_t y_t$; thus, there is no need to explicitly define the set S' in the inequality):

$$\sum_{t \in S} x_t \leq d_{u,l} - \sum_{t \in S}(k_t - \lambda)^+(1 - y_t) + i_l. \tag{6.29}$$

Let us next consider a set of periods that are in L but are not in the set S, and let $N \subseteq L \backslash S$ denote such a subset of periods. The validity of (6.29) and $x_t \leq k_t y_t$ for all $t = 1, \ldots, T$, imply the validity of the inequality

$$\sum_{t \in S \cup N} x_t \leq d_{u,l} - \sum_{t \in S}(k_t - \lambda)^+(1 - y_t) + \sum_{t \in N} k_t y_t + i_l. \tag{6.30}$$

We can strengthen inequality (6.30) by reducing the coefficients of the y_t variables for some or all of the variables in the set N, without destroying the validity of the inequality. That is, suppose we would like to determine a coefficient α_t for each $t \in N$ such that

$$\sum_{t \in S \cup N} x_t + \sum_{t \in S}(k_t - \lambda)^+(1 - y_t) - \sum_{t \in N} \alpha_t y_t \leq d_{u,l} + i_l \tag{6.31}$$

is a valid inequality. To analyze this inequality, let S^+ denote the set of $t \in S$ such that $k_t > \lambda$, let S_0 (N_0) denote the set of $t \in S$ ($t \in N$) such that $y_t = 0$, and let $S_1 = S \backslash S_0$ and $N_1 = N \backslash N_0$. Then we can write this inequality as

$$\sum_{t \in S_1 \cup N_1} x_t + \sum_{t \in S^+ \cap S_0} (k_t - \lambda) - \sum_{t \in N_1} \alpha_t \leq d_{u,l} + i_l. \tag{6.32}$$

Suppose we would like to ensure that the value of α_t for $t \in N_1$ offsets the increase in the value of the left-hand side of (6.32) for any $t \in S^+ \cap S_0$ (i.e., $k_t - \lambda$). To this end, let $k_S = \max_{t \in S}\{k_t\}$, let $\bar{k}_t = \max\{k_S; k_t\}$, and suppose we let $\alpha_t = \bar{k}_t - \lambda$ for all $t \in N$. Then (6.32) becomes

$$\sum_{t \in S_1 \cup N_1} x_t + \sum_{t \in S^+ \cap S_0} (k_t - \lambda) - \sum_{t \in N_1} (\bar{k}_t - \lambda) \leq d_{u,l} + i_l. \tag{6.33}$$

This inequality is valid as long as the left-hand side is less than or equal to $d_{u,l} + i_l$ for all choices of the sets N_1, S_1, and S_0. Clearly, if $|N_1| \geq |S^+ \cap S_0|$, the left-hand side is less than or equal to $\sum_{t \in S_1 \cup N_1} x_t$, which must be less than or equal to $d_{u,l} + i_l$. Suppose instead that $|N_1| < |S^+ \cap S_0|$. Then the left-hand side of (6.33) is less than or equal to $\sum_{t \in S_1 \cup N_1} k_t + \sum_{t \in S^+ \cap S_0} k_t - |S^+ \cap S_0|\lambda - \sum_{t \in N_1} \bar{k}_t + |N_1|\lambda$, which is less than or equal to $\sum_{t \in S} k_t - \lambda + \sum_{t \in N_1}(k_t - \bar{k}_t) + \lambda(|N_1| + 1 - |S^+ \cap S_0|)$. Because $d_{u,l} = \sum_{t \in S} k_t - \lambda$, $k_t \leq \bar{k}_t$ for all $t \in L$, and $|N_1| < |S^+ \cap S_0|$, this quantity must be less than or equal to $d_{u,l} + i_l$. Therefore, the following inequality is valid for the CELSP:

$$\sum_{t \in S \cup N} x_t + \sum_{t \in S}(k_t - \lambda)^+(1 - y_t) - \sum_{t \in N}(\bar{k}_t - \lambda)y_t \leq d_{u,l} + i_l. \tag{6.34}$$

Noting, however, that period $t \in N$ can never contribute more than k_t units of capacity, the resulting generalized flow-cover inequality can be written as[3]

$$\sum_{t \in S \cup N} x_t \leq d_{u,l} - \sum_{t \in S}(k_t - \lambda)^+(1 - y_t) + \sum_{t \in N} \min\{\bar{k}_t - \lambda; k_t\}y_t + i_l. \tag{6.35}$$

Padberg, Van Roy, and Wolsey [89] show that this inequality is valid for a polyhedron corresponding to a flow problem with variable upper bounds (in which class the CELSP falls). Their form of the inequality uses the coefficient $(\bar{k}_t - \lambda)$ for y_t, $t \in N$, as in (6.34), noting that the inequality is only useful when $k_t > \bar{k}_t - \lambda$ for $t \in N$, as otherwise the inequality is dominated by a combination of the inequality (6.35) when t is removed from N, and the inequality $x_t \leq k_t y_t$.

Pochet [93] specialized the generalized flow-cover inequalities to the CELSP and demonstrated conditions under which they serve as facets for the convex hull of mixed integer solutions for the equal-capacity case. In doing so, he recognized that the effective capacity of some period $t \in L$ for satisfying demand in periods u through l is equal to the minimum between k_t and $d_{t,l}$. Thus, he effectively defines the capacity of a period $t \in L$ as $\bar{d}_{t,l} = \min\{k_t, d_{t,l}\}$, as well as $\bar{k} = \max_{t \in S}\{\bar{d}_{t,l}\}$ and $\tilde{d}_{t,l} = \max\{\bar{k}, \bar{d}_{t,l}\}$ (here $\tilde{d}_{t,l}$ plays the role of $\bar{k}_t$ in (6.35)). Using these definitions of effective capacity, and assuming $u \in S$, Pochet provides the following form of the generalized flow-cover inequalities for the CELSP:

$$\sum_{t \in S \cup N} x_t \leq d_{u,l} - \sum_{t \in S}(\bar{d}_{t,l} - \lambda)^+(1 - y_t) + \sum_{t \in N} \min\{\tilde{d}_{t,l} - \lambda; \bar{d}_{t,l}\}y_t + i_l. \tag{6.36}$$

As we will see in the next chapter, the generalized flow-cover inequalities also play a role in strengthening the LP relaxation for discrete facility location problems.

[3]Note that if we had defined N as a subset of periods in $L \backslash S$ such that $k_t > \bar{k}_t - \lambda$, then inequality (6.35) can be obtained by aggregating inequality (6.34) with valid inequalities of the form $x_t \leq k_t y_t$ for a set of variables $N' \in L \backslash \{S \cup N\}$ such that $k_t \leq \bar{k}_t - \lambda$, and redefining the set N as equal to $N \cup N'$.

6.7 Review

This chapter generalizes the previous chapter to account for the practical realities of limited production capacities. The remarkable paper by Florian and Klein [49] appears to have been the starting point for the many contributions and developments for this problem class over the past forty-plus years. This paper provided both a characterization of extreme point solutions and a powerful insight that led to a polynomial time algorithm for the equal-capacity case (van Hoesel and Wagelmans [111] later improved the worst-case complexity for solving this problem by an order of magnitude). The paper by Florian et al. [50] provides the connection to the knapsack problem discussed in Section 6.3 as well as an extremely clever proof of $\mathcal{NP}$-Hardness for the CELSP. Van Hoesel and Wagelmans [112] provided the powerful results and novel dynamic programming formulation discussed in Section 6.5, which lead to a fully polynomial time approximation scheme for the CELSP. The polyhedral results in Section 6.6 originated with the study of the convex hull of solutions for the UELSP by Barany et al. [12], while the inequalities for the CELSP originated with the study of flow problems with variable upper bounds by Padberg et al. [89]. Leung et al. [75] and Pochet [93] provided the key inequalities for the CELSP and demonstrated conditions under which they serve as facets of the convex hull of mixed integer solutions for the equal-capacity CELSP.

Exercises

Ex. 6.1 — Consider the LP relaxation of the CELSP formulation (6.1)–(6.5) and that of formulation UELSP″ (5.13)–(5.16) with the additional capacity constraints $\sum_{\tau=t}^{T} d_\tau x_{t,\tau} \leq k_t y_t$, for $t = 1, \ldots, T$. Show that the former relaxation admits solutions that the latter does not, while the converse does not hold (implying that the LP relaxation of the latter formulation provides a lower bound at least as high as that of the former).

Ex. 6.2 — Extend the formulation of the CELSP in Section 6.2 to the case in which backordering is permitted in each period at a cost of b_t for each unit backordered at the end of period t, $t = 1, \ldots, T$.

Ex. 6.3 — How would you characterize a regeneration interval (RI) for the CELSP in the case that backordering is permitted?

Ex. 6.4 — How would Property 1 and Corollary 1 change if backordering is permitted?

Ex. 6.5 — For the equal-capacity CELSP shortest path network in Example 6.2, shown in Figure 6.2, compute the cost of each arc in the network.

7

Multistage Production and Distribution Planning Problems

7.1 Introduction

Consumer products often consist of multiple components and subassemblies that require assembly before a product takes its final form. The needed components and subassemblies may require the use of various technologies, equipment, and operations located at different stages and physical locations. Associated with a complex product is typically an overall recipe for creating the product, which encodes the necessary components, component and subassembly order lead times, and sequence of operations, and is known as a *bill-of-materials*. The order lead times arise between different production stages and correspond to the time between when a downstream stage places an order with an upstream stage, and the time the order is received by the downstream stage and ready for use in subsequent operations.

It is convenient to represent the various production and distribution stages as nodes in a network, with directed arcs connecting the nodes (stages). Figure 7.1 illustrates different kinds of network structures that may arise in production and distribution systems. A *serial system* has a linear network structure in which each node has at most one incoming arc and at most one outgoing arc, i.e., at most one immediate predecessor and at most one immediate successor. Serial systems are often found in manufacturing flow lines within a plant. In an *assembly system* each node has at most one immediate successor, but may have multiple immediate predecessors. Such structures often arise in the assembly of a product with a complex bill-of-materials, where the final stage (the only stage with no successors) corresponds to the final product. In a *distribution system*, in contrast, any node may have multiple immediate successors, but may have at most one immediate predecessor. As the name indicates, these structures are often found in the distribution of final products to markets via a network of warehouses. Serial, assembly, and distribution system networks are not only necessarily acyclic, but they also possess the property that at most one path exists between any pair of nodes (note also that a serial system is a special case of both the assembly and distribution systems). In a *general system*, on the other hand, a node may have both multiple immediate predecessors and multiple immediate successors, which does not preclude the existence of cycles in the network, although we assume that no directed cycles exist (i.e., no directed path exists that leaves from and returns to the same node). Systems in which the final stage of an assembly system serves as the initial stage in a distribution system are not uncommon in practice. Such a network would be classified as a general system, despite containing no cycles.

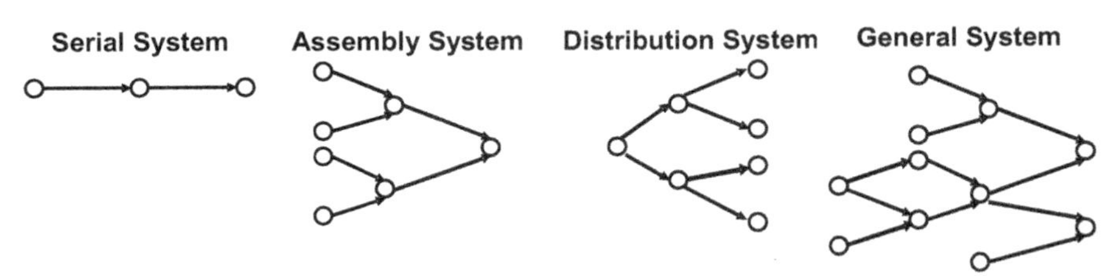

FIGURE 7.1
Network structures representing production and distribution system operations.

In the analysis of multistage production and distribution systems, the ownership and objectives of different stages have a substantial influence on the system's operation, as well as on our analysis of the system. When stages have different ownership and objectives we call this a decentralized system. Management of decentralized systems serves as a focal point of research in supply chain management, where the objectives of decision makers at different stages may lead to conflicting incentives, often requiring the application of game-theoretic analysis (see, e.g., Chopra and Meindl [24]). Our analysis of multistage lot sizing problems, in contrast, will focus on centralized systems, in which a single decision maker determines the timing and quantities involved in the movement of goods between stages. The analysis of optimal solutions in centralized systems also plays an important role in characterizing the performance of decentralized systems, as the minimum total system cost provides a lower bound on the cost of a decentralized system.

We confine our analysis throughout this book to deterministic systems, in which all cost, constraint, and demand parameters are known with certainty. Such analyses are more appropriate for systems in which production and distribution occurs based on a predetermined plan (e.g., so-called *push* systems), or when production/distribution activities occur in response to customer demands, i.e., *make-to-order* systems. Section 7.2 considers discrete-time models in which demands, costs, and constraint parameters may vary from time period to time period. These models correspond to multistage generalizations of the economic lot sizing problems considered in Chapters 5 and 6. Section 7.3 then considers multistage systems in which external demands occur at a continuous, constant rate and the problem's cost and demand parameters are time invariant. The resulting continuous-time models correspond to multistage generalizations of the economic order quantity (EOQ) model ([65]).

7.2 Models with Dynamic Demand

This section considers discrete-time models with dynamic demand and cost parameters, as in Chapters 5 and 6. In a multistage system with dynamic demands and costs, all variables and parameters must have associated stage indices in addition to time indices (with the possible exception of the demand parameters in serial and assembly systems; if demand only occurs at a final stage, then the stage index for demand is unnecessary). Although the results we will discuss are valid for general concave cost functions of production and inventory levels, as in Chapters 5 and 6, we will consider the special case of concave cost functions consisting of fixed plus linear variable production costs and linear holding costs. Thus, for example, f_{jt}, c_{jt}, and h'_{jt} will denote the fixed production/order cost, unit production cost, and unit holding cost at stage j in period t, for $j \in J$ and $t = 1, \ldots, T$, where J denotes the set of production stages, and T denotes the number of periods in the planning horizon. We continue to apply the convention of charging holding cost against end-of-period inventory, where i'_{jt} denotes the local end-of-period t inventory at stage j. The variables x_{jt} and y_{jt}, respectively, correspond to the production level and the binary decision variable associated with a production setup (or replenishment order) at stage j in period t (we will refer to any type of replenishment order as a production setup). We let d_{jt} denote the demand at stage j in period t (if demand occurs only at a single stage then we suppress the stage index for demands).

7.2.1 Serial systems with dynamic demand

Consider an n-stage serial system in which stage j must meet the demands of stage $j+1$, external demand occurs only at stage n, and stage 1 replenishes its inventory from an external source with unlimited capacity. We assume in this subsection that production output in any period at any stage is unlimited, backorders are not permitted at any stage in the system, and the replenishment lead time at any stage is finite and constant.[1] The formulation of the production planning problem for this serial system is a straightforward generalization of the UELSP in Chapter 5. An inventory balance constraint exists for each stage, where the downstream production levels at stage $j + 1$ serve as the requirements at stage j, with the exception of stage n, which faces requirements equal to the external demands $d_1, \ldots, d_T$. As in the UELSP, a production forcing constraint exists at each stage as well, such that positive production is permitted in period t at stage j if and only if a production setup occurs at the stage in the period (i.e., if $y_{jt} = 1$). The resulting formulation

[1]Given a system with finite and constant replenishment lead times at each stage, an equivalent problem instance exists in which all lead times are zero.

of the serial uncapacitated lot sizing (SULS) problem is written as follows.

[SULS]

$$\text{Minimize} \quad \sum_{t=1}^{T} \sum_{j \in J} \{f_{jt}y_{jt} + c_{jt}x_{jt} + h'_{jt}i'_{jt}\} \tag{7.1}$$

$$\text{Subject to:} \quad i'_{j,t-1} + x_{jt} = x_{j+1,t} + i'_{jt}, \quad j \in J \backslash \{n\}, \; t = 1, \ldots, T, \tag{7.2}$$

$$i'_{n,t-1} + x_{nt} = d_t + i'_{nt}, \quad t = 1, \ldots, T, \tag{7.3}$$

$$x_{jt} \leq M_t y_{jt}, \quad j \in J, t = 1, \ldots, T, \tag{7.4}$$

$$i'_{j0} = 0, x_{jt}, i'_{jt} \geq 0, \quad j \in J, t = 1, \ldots, T, \tag{7.5}$$

$$y_{jt} \in \{0, 1\}, \quad j \in J, t = 1, \ldots, T. \tag{7.6}$$

The SULS problem can be represented as a fixed-charge network flow problem as shown in Figure 7.2 for a problem with three stages and six time periods. As in the UESLP, a single source node exists in the network with supply equal to the sum of all of the demands, where $D(s,t) = \sum_{\tau=s}^{\tau=t} d_\tau$. This network has a node for each stage-period pair; the six nodes in the top layer directly below the source correspond to stage zero (the external supplier stage) in the six time periods, and the arcs from left-to-right connecting nodes within this layer have zero cost. These nodes feed stage one production, which feeds stage one inventory and stage two production, and so on.

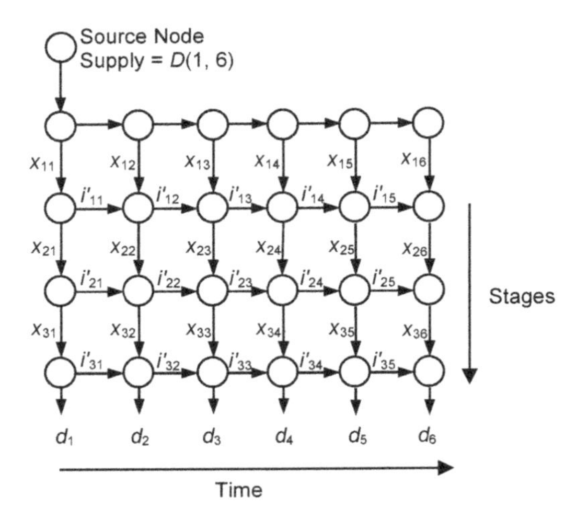

FIGURE 7.2
Fixed-charge network flow representation of the serial uncapacitated lot sizing (SULS) problem for a problem instance with three stages and six time periods.

As in the UELSP, because the objective function is a concave function of the production and inventory levels, and the constraints form a linear feasible region, an optimal extreme point solution exists in the network illustrated in Figure 7.2. An extreme point solution in the network corresponds to a solution in which the arcs corresponding to basic variables (i.e., all arcs with positive flow) form a spanning tree. This implies that an optimal solution exists satisfying the zero-inventory ordering (ZIO) property defined in Section 5.2.1 at each stage and in every time period. Thus, we only need to consider solutions such that production at a stage in a time period implies that no inventory was held at the stage at the end of the previous period (and vice versa, i.e., inventory held at the end of a period implies we do not need to consider producing in the subsequent period). This immediately implies that we may limit our consideration to solutions in which the production quantity at any stage in any period t equals $D(t_1, t_2)$ for some $t \leq t_1 \leq t_2$.

As Zangwill [118] showed, this permits a polynomial time solution algorithm for the SULS problem via dynamic programming. To illustrate this, consider the network in Figure 7.2. Associated with each node in the network is a stage-period pair, (j, t), for stages 0 through n and periods 1 through T, where stage 0 corresponds to the external supplier (for example, the bottom right node in the figure corresponds to node $(3, 6)$, while the top right node corresponds to $(0, 6)$). The total flow exiting node (j, t) corresponds to output that has completed processing at stage j in period t and is either held in inventory at stage j at the end of period t (this amount equals i'_{jt}) or moves to production at stage $j+1$ in period t (this amount equals $x_{j+1,t}$; "production" at stage $n+1$ corresponds to demand, and has an associated production cost of zero). In an extreme point solution, not only must the sum of these values $(x_{j+1,t} + i'_{jt})$ equal $D(t_1, t_2)$ for some $t_2 \geq t_1 \geq t$, but we must also have $x_{j+1,t} = D(t_1, \tau)$ while $i'_{jt} = D(\tau + 1, t_2)$ for some $t_1 - 1 \leq \tau \leq t_2$ (where $D(t, t-1) = 0$). Note that flow from node $(0, t)$ to $(1, t)$ corresponds to production in period t at stage 1, while flow from node $(0, t)$ to $(0, t+1)$ occurs at zero cost, and the dummy source node sends flow only to node $(0, 1)$.

With this network in mind, we define $C_{jt}(t_1, t_2)$ as the minimum cost, after stage j production in period t, at stages j through n in periods t through t_2, assuming exactly $D(t_1, t_2)$ units of flow exit node (j, t), $t_2 \geq t_1 \geq t$. With this definition, the optimal solution takes a value equal to $C_{01}(1, T)$. We then have

$$C_{jt}(t_1, t_2) = \min_{\tau = t_1 - 1, \ldots, t_2} \{(f_{j+1,t} + c_{j+1,t} D(t_1, \tau) + C_{j+1,t}(t_1, \tau))\delta_{\{\tau \geq t_1\}} + h'_{jt} D(\tau + 1, t_2) + C_{j,t+1}(\tau + 1, t_2)\}, \quad (7.7)$$

where $D(t, t-1) = 0$ for all values of t, and $\delta_{\{\tau \geq t_1\}} = 1$ if $\tau \geq t_1$, and equals zero otherwise. The first term in the minimization in (7.7) gives the cost at stages $j+1$ through n in periods t through τ, associated with satisfying the demand in periods t_1 through τ (this term is only valid, therefore, if $\tau \geq t_1$). The second term gives the cost, after production at stage j in period t, at stages j through n in periods t through t_2, associated with satisfying the demand in periods $\tau + 1$ through t_2. We set the boundary conditions

$C_{n+1,t}(t,t) = 0$, while $C_{n+1,t}(t,\tau) = \infty$ if $\tau \neq t$ and $C_{j,T+1}(T,T) = \infty$ for all $j = 1,\ldots,n$ (with $C_{j,T+1}(T+1,T) = 0$ for all $j = 1,\ldots,n$). With this setup, we begin by computing $C_{jT}(T,T)$ for $j = 1,\ldots,n$, and work backwards in time and in the stages. For each value of j and t, we have $\mathcal{O}(T^2)$ pairs (t_1,t_2), and evaluating each pair requires considering $\mathcal{O}(T)$ values of τ; thus, for each j and t pair, we require $\mathcal{O}(T^3)$ operations. This implies a worst-case complexity of $\mathcal{O}(nT^4)$.

Observe that the dynamic programming recursion (7.7) may produce an optimal solution in which all production at a stage in some period t is used to satisfy demand in periods t_1 through t_2, with $t_1 > t$. In such a solution, it must be the case that all of the production at some stage j in some period t is held in inventory at the end of period t, i.e., production occurs solely for stock. Under certain structural cost assumptions, however, it is possible to show that an optimal solution exists in which production does not occur solely for stock at any stage in any period. Love [81] showed that this is the case under nonincreasing production costs over time and nondecreasing holding costs in the stage numbers. In other words, if $f_{jt} + c_{jt}x \geq f_{j,t+1} + c_{j,t+1}x$ for all $x \geq 0$, each stage $j = 1,\ldots,n$, and $t = 1,\ldots,T-1$, and $h'_{jt} \leq h'_{j+1,t}$ for all $j = 1,\ldots,n-1$ and $t = 1,\ldots,T$, then an optimal solution exists in which production does not contribute solely to stock at any stage in any period.[2] This implies that if an optimal solution exists such that an order is placed at some stage $j < n$ in period t, then an optimal solution exists such that an order is placed at stage $j+1$ in period t. Under these structural cost assumptions, therefore, we can confine our attention to solutions such that an order placed at stage j in period t implies that an order is placed at stages $j+1$ through n in period t.

A solution of this type is consistent with the definition of a *nested policy* in the inventory literature. A nested policy requires that when a stage orders in a period, all successor stages also order in the period; under the non-speculative cost assumptions proposed by Love [81], a nested policy is optimal (Exercise 7.1 asks the reader to prove this). This implies that if a stage produces in period t, then it produces at least d_t units, as do all successor stages. This further implies that in computing $C_{jt}(t_1,t_2)$, we need not consider values of $t_1 > t$. Thus, we can redefine the value function as $C_{jt}(t,t_2)$ for all $j = 1,\ldots,n$, $t = 1,\ldots,T$, and $t_2 \geq t$. For a given value of t and stage j, we compute $C_{jt}(t,t_2)$ for $\mathcal{O}(T)$ values of t_2 and $\mathcal{O}(T)$ values of τ. The resulting worst-case complexity can thus be characterized as $\mathcal{O}(nT^3)$.

[2] Production and holding costs that violate these conditions are often said to imply speculative motives in the stochastic inventory literature. Speculative motives lead to production solely for stock at a stage in order to meet demand in a later period. These conditions are, therefore, consistent with the concept of non-speculative motives, implying that no incentive exists to produce in a period and hold all of this production in inventory to meet demand in a later period. This also implies that an optimal solution exists such that demand at a stage in a period will always use the most recent production setup at the prior stage in order to meet demand at the stage.

7.2.2 Production networks with non-speculative costs

We next discuss dynamic versions of multistage production and distribution problems with general network structures, in which nodes in the corresponding network may have multiple predecessors and/or successors. We begin with the case of assembly networks, where each node has at most one successor, but may have multiple predecessors. As in the previous section on serial systems, we assume that demand in the assembly system occurs only at a final stage n, which is the sole stage in the network with no successors. We employ the same notation defined for serial systems in the previous subsection, with some additional notation to account for the precedence structure in the graph. For any node $j \in J \backslash \{n\}$, let $s(j)$ denote the unique immediate successor of node j, and let $\mathcal{P}(j)$ denote its set of all immediate predecessors. We take a more general view of the interactions between stages than described in the serial system case, assuming that a unit produced at node i requires α_{ji} units of production at stage j if $i = s(j)$. Our discussion of assembly systems will focus on the problem considered by Crowston and Wagner [34] and Afentakis, Gavish, and Karmarkar [4], assuming unlimited capacity at each stage, nonincreasing stage production costs in time, and that no shortages are permitted (these cost assumptions are sometimes referred to as non-speculative costs, as discussed in Footnote 2 on Page 100).

Before presenting the assembly problem formulation, we introduce the concepts of *echelon inventory* and *echelon holding cost*, which will serve as useful constructs for formulating and solving multistage inventory problems. The echelon inventory at a stage consists of the local inventory at the stage, plus the sum of the local inventory levels at all successor stages, where a successor stage is any node reachable on a directed path in the network representation of the system (as in, for example, Figure 7.1). The echelon unit holding cost at a stage equals the local holding cost of a unit minus the sum of the equivalent unit holding costs at all immediate predecessor stages (we will subsequently define the echelon holding cost more precisely). Multiplying the echelon holding cost by the echelon inventory level at each stage (and summing over all stages) provides a mechanism for computing inventory holding costs that is equivalent to multiplying the local inventory holding cost by the local inventory level at each stage (and summing over all stages; Exercise 7.3 asks the reader to show this for a two-stage serial system). Because the value of inventory typically increases at downstream stages in practice, we assume that all echelon inventory costs are nonnegative.

The problem formulation we describe for uncapacitated assembly systems uses an echelon inventory formulation, which requires keeping track of the nodes on the *path* $\pi(j)$ from node j to node n, for each $j \in J \backslash \{n\}$. Note that this path contains the set of all successors of node j. Observe that the number of units of output at stage j required to make one unit of output at stage i for $i \in \pi(j)$, denoted as ρ_{ji}, equals

$$\rho_{ji} = \prod_{k \in \pi(j) \backslash \pi(i)} \alpha_{k,s(k)}. \tag{7.8}$$

The echelon inventory at stage j at the end of period t is characterized by

$$i_{jt} = i'_{jt} + \sum_{i \in \pi(j)} \rho_{ji} i'_{it}, \tag{7.9}$$

while the echelon holding cost rate at stage j is given by

$$h_{jt} = h'_{jt} - \sum_{i \in \mathcal{P}(j)} \alpha_{ij} h_{it}. \tag{7.10}$$

The echelon formulation of the uncapacitated assembly system (UAS) lot sizing problem can then be written as

[UAS]

$$\begin{aligned}
\text{Minimize} \quad & \sum_{t=1}^{T} \sum_{j \in J} \{f_{jt} y_{jt} + c_{jt} x_{jt} + h_{jt} i_{jt}\} & & (7.11)\\
\text{Subject to:} \quad & i_{j,t-1} + x_{jt} = \rho_{jn} d_t + i_{jt}, & j \in J, t = 1, \ldots, T, & (7.12)\\
& i_{jt} \geq \alpha_{j,s(j)} i_{s(j),t} & j \in J \backslash \{n\}, & (7.13)\\
& & t = 1, \ldots, T, & \\
& x_{jt} \leq M_t y_{jt}, & j \in J, t = 1, \ldots, T, & (7.14)\\
& i_{j0} = 0, x_{jt}, i_{nt} \geq 0, & j \in J, t = 1, \ldots, T, & (7.15)\\
& y_{jt} \in \{0, 1\}, & j \in J, t = 1, \ldots, T. & (7.16)
\end{aligned}$$

Note that the use of echelon inventory variables permits writing the demand for echelon inventory at stage j in period t as $\rho_{jn} d_t$ in the inventory balance constraints (7.12). Constraint set (7.13) ensures nonnegative installation inventory levels. Note that because the production costs are nonincreasing in time, and assuming echelon inventory holding cost rates are nonnegative, then just as in the serial system case, nested production schedules are optimal. In addition, the concavity of the objective function and the linearity of the feasible region imply that a zero-inventory ordering (ZIO) solution is optimal, i.e., $x_{jt} i'_{j,t-1} = 0$ for all $j \in J$ and $t = 1, \ldots, T$. A nested policy together with a zero-inventory ordering policy implies that a zero-echelon-inventory ordering (ZEIO) policy is optimal as well, i.e., $x_{jt} i_{j,t-1} = 0$ for all $j \in J$ and $t = 1, \ldots, T$.

Crowston and Wagner [34] provide a dynamic-programming-based algorithm for solving problem UAS, although the running time of this algorithm increases exponentially in T (they also provide a customized branch-and-bound algorithm). Afentakis et al. [4] provided a Lagrangian-relaxation-based solution approach that enabled solving much larger problem sizes. This scheme relaxes constraint set (7.13) using a Lagrangian multiplier μ_{jt} for each $j \in J \backslash \{n\}$ and $t = 1, \ldots, T$. Letting $h^{\mu}_{jt} = h_{jt} + \sum_{i \in \mathcal{P}(i)} \alpha_{ij} \mu_{it} - \mu_{jt}$ the resulting Lagrangian relaxation formulation is written as follows.

[UASLR]

$$\text{Minimize} \quad \sum_{t=1}^{T}\sum_{j\in J}\left\{f_{jt}y_{jt}+c_{jt}x_{jt}+h^{\mu}_{jt}i_{jt}\right\} \tag{7.17}$$

$$\text{Subject to:} \quad i_{j,t-1}+x_{jt}=\rho_{jn}d_t+i_{jt}, \quad j\in J, t=1,\ldots,T, \tag{7.18}$$

$$x_{jt}\le M_t y_{jt}, \quad j\in J, t=1,\ldots,T, \tag{7.19}$$

$$i_{j0}=0, x_{jt}, i_{nt}\ge 0, \quad j\in J, t=1,\ldots,T, \tag{7.20}$$

$$y_{jt}\in\{0,1\}, \quad j\in J, t=1,\ldots,T. \tag{7.21}$$

Problem UASLR decomposes by stage, resulting in a single subproblem for each stage $j \in J$. For a fixed vector of Lagrangian multipliers, μ, this subproblem, for a given $j \in J$, looks like a single-stage UELSP; note, however, that it is possible that $h^{\mu}_{jt} < 0$, or equivalently, we may have negative holding costs. In addition, the resulting subproblem permits negative inventory levels. Note, however, that we can add valid inequalities to the original formulation UAS enforcing nonnegative echelon inventory levels and ZEIO conditions at each stage in every period. That is, if we add $i_{jt} \ge 0$ and $x_{jt}i_{j,t-1} = 0$ for all $j \in J$ and $t = 1, \ldots, T$ to the UAS formulation (for which these constraints are redundant), they are inherited by the UASLR formulation, where they are not necessarily redundant. This implies that we can use the UELSP shortest path solution method for solving each resulting subproblem for a fixed Lagrangian multiplier vector μ.

Afentakis et al. [4] observed extremely tight bounds on the optimal solution value via this Lagrangian relaxation approach for problems with time-invariant costs. Their computational tests included problem instances with between 5 and 50 nodes and between 6 and 18 time periods (120 problems in total). The worst lower bound they observed upon maximizing the Lagrangian dual problem (via subgradient optimization) was within 0.46% of the optimal solution value. They used these tight bounds within a customized branch-and-bound algorithm, permitting them to solve most of their problems within a minute (only four of the 18-period, 50-node problems reached the time limit of 270 seconds, and the best solutions found for these instances were within 5×10^{-4} of the optimal solution).

Afentakis and Gavish [3] subsequently considered more general acyclic network structures containing a set J of n nodes, or stages, where $\mathcal{S}(j)$ denotes the set of immediate successors of node j. Because a node may have multiple paths in the network to another node i, we now let π_{ji} denote the set of paths from node j to node i (where each path corresponds to a set of nodes). A path $p \in \pi_{ji}$ consists of a set of $l(p)$ nodes $j = j_{1(p)}, j_{2(p)}, \ldots, j_{l(p)} = i$. The number of units of stage j production used in the production of one unit at stage i, denoted as ρ_{ji}, is given by

$$\rho_{ji} = \sum_{p\in\pi_{ji}} \prod_{k=1}^{l(p)-1} \alpha_{j_{k(p)},j_{k+1(p)}}. \tag{7.22}$$

As in the UAS formulation, we assume that each stage incurs fixed plus per-unit production costs, as well as unit holding costs. The only modification needed to transform the UAS formulation to account for this more general case requires changing the echelon inventory constraint set (7.13) to

$$i_{jt} \geq \sum_{i \in \mathcal{S}(j)} \alpha_{ji} i_{it}, \quad j \in J \backslash \{n\}, t = 1, \ldots, T. \tag{7.23}$$

As in the assembly case, we know that an optimal ZIO solution exists. Under non-speculative costs, Afentakis and Gavish [3] showed that an optimal solution exists that satisfies a form of the nestedness property as well. That is, if an optimal solution exists with $x_{jt} > 0$, then an optimal solution exists such that for some $i \in \mathcal{S}(j)$, $x_{it} > 0$ as well. In addition, Afentakis and Gavish [3] showed that if node j and all of its successors produce in period t, then $i_{j,t-1} = 0$. They then showed that the general network problem we have described can be transformed into an equivalent problem with an assembly structure. To do this, given any node $j \in J \backslash \{n\}$, we create $|\pi_{jn}| - 1$ copies of the node (in addition to the original node). Any node in the new network is denoted as j^p, indicating the instance of node j associated with path $p \in \pi_{jn}$, and an arc with weight α_{ji} extends from node j^p to i^p in the new network if and only if $i \in \mathcal{S}(j)$. Note that the resulting network contains an assembly structure. We cannot simply apply the UAS formulation to this new network, however, without adding some new constraints and modifying the objective function slightly. The new constraints must account for the fact that if a setup occurs at some node j^1, for example, then production may occur at all nodes j^p, $p \in \pi_{jn}$, without incurring additional setup costs. To account for these constraints, for any node with multiple paths to node n, let $\tilde{y}_{jt}$ denote a binary variable equal to one if production occurs at any node j^p, $p \in \pi_{jn}$, in period t. Then, letting J^1 denote the set of nodes with a single path to node n (including node n itself), we add the new constraint set

$$y_{j^p t} \leq \tilde{y}_{jt}, \quad j^p \in J \backslash J^1, t = 1, \ldots, T. \tag{7.24}$$

The objective function term associated with the fixed costs incurred in production is now equal to $\sum_{t=1}^{T} (\sum_{j \in J \backslash J^1} f_{jt} \tilde{y}_{jt} + \sum_{j \in J^1} f_{jt} y_{jt})$, which replaces the original term $\sum_{t=1}^{T} \sum_{j \in J} f_{jt} y_{jt}$.

When we relax the echelon inventory constraints (7.13) in this modified assembly problem formulation, constraint set (7.24) implies that the resulting subproblems do not completely decompose into a set of independent single-node subproblems. Instead, a subproblem exists that is associated with each of the original network's stages, but each such subproblem may involve joint (or shared) replenishment costs. This subproblem is, in general, $\mathcal{NP}$-Hard. Thus, solving the resulting Lagrangian relaxation problem will not involve subproblems that we can solve efficiently. If we also relax the new constraint set (7.24), using a Lagrangian multiplier $\theta_{j^p t}$ for each (j^p, t) pair such that $j \in J \backslash J^1$, the problem decomposes into a set of $|\pi_{jn}|$ subproblems for each

$j \in J \backslash J^1$, plus one subproblem for each stage $j \in J^1$, plus one additional subproblem. The subproblem associated with each node in J^1 takes the form of UASLR. A total of $|\pi_{jn}|$ subproblems are also created for each node $j \in J \backslash J^1$, such that the subproblem associated with node j^p takes the form of UASLR, except that the fixed cost term associated with node j^p equals the Lagrangian multiplier $\theta_{j^p t}$. Each of the subproblems we have discussed thus far corresponds to a single-stage lot sizing problem that may be solved via a shortest path algorithm, just as in the case of an assembly network structure.

One additional subproblem also arises, which takes the following form.

$$\text{Minimize:} \quad \sum_{j \in J \backslash J^1} \sum_{t=1}^{T} \left(f_{jt} - \sum_{k=1}^{|\pi_{jn}|} \theta_{j^k t} \right) \tilde{y}_{jt} \tag{7.25}$$

$$\text{Subject to:} \quad \tilde{y}_{jt} \in \{0, 1\}, \qquad j \in J \backslash J^1. \tag{7.26}$$

For a fixed vector θ, this problem is easily solved by inspection. We set $\tilde{y}_{jt} = 1$ if $f_{jt} \leq \sum_{k=1}^{|\pi_{jn}|} \theta_{j^k t}$, and $\tilde{y}_{jt} = 0$ otherwise.

Afentakis and Gavish [3] implemented a subgradient optimization algorithm for maximizing the Lagrangian dual, and used this to provide bounds for a customized branch-and-bound algorithm. They again showed that this Lagrangian relaxation approach provides very sharp bounds on the optimal solution value, even for this more general class of networks.

7.2.3 Constant-factor approximations for special cases

We next discuss some specialized network structures that allow approximation algorithms with constant-factor worst-case performance guarantees under dynamic demands and costs. The first such problem class we consider corresponds to the dynamic version of the one-warehouse multi-retailer (OWMR) problem. For this problem, we have a set J of n retailers, each of whom is served by a warehouse; in the corresponding network, node j corresponds to retailer j for $j > 0$, while node 0 corresponds to the warehouse. An arc extends from node zero to each node $j \in J$; thus, the resulting network corresponds to a two-level distribution system. Let d_{jt} and f_j denote the demand and fixed order cost at retailer j in period t for each $j \in J$ and $t = 1, \ldots, T$, while f_{0t} denotes the fixed order cost at the warehouse in period t for $t = 1, \ldots, T$. We restrict ourselves to time-invariant fixed order costs at the retailers. As Chan, Muriel, Shen, Simchi-Levi, and Teo [22] point out, otherwise the problem is equivalent in difficulty to a set covering problem, a problem for which a constant-factor approximation algorithm is unlikely to exist (unless $\mathcal{P} = \mathcal{NP}$).

Levi, Roundy, Shmoys, and Sviridenko [77] provided a constant-factor approximation algorithm for the dynamic OWMR problem, based on an LP relaxation rounding scheme. Let h^j_{rst} denote the cost to hold a unit of inventory at the warehouse from period r to period $s-1$, and then at retailer j from period s to period t ($r \leq s \leq t$, assuming shortages are not permitted). As with our prior analysis of dynamic planning problems with fixed order costs and linear procurement and holding costs, an optimal ZIO solution exists for the OWMR problem, both at the warehouse level and the retailer level. Because of this, we can consider solutions such that all of the demand at retailer $j \in J$ in period t is ordered in some period $s \leq t$, and such that the corresponding inventory is delivered to the warehouse in some period $r \leq s$. Let x^j_{rst} denote the percentage of retailer j demand in period t that is received at the warehouse in period r, shipped to retailer j in period s, and used to satisfy demand in period t (for each $j \in J$ and all (r, s, t) such that $1 \leq r \leq s \leq t \leq T$). The ZIO property implies that an optimal solution exists such that these x^j_{rst} variables equal zero or one at optimality. We thus define $H^j_{rst} = h^j_{rst} d_{jt}$ as the total cost to satisfy demand at retailer j in period t, assuming that the corresponding inventory arrives at the warehouse in period r and is shipped to the retailer in period s (for all (r, s, t) such that $1 \leq r \leq s \leq t \leq T$). Let $y_{jt} = 1$ ($y_{0t} = 1$) if retailer j (the warehouse) places an order in period t. Then we can formulate the dynamic one-warehouse multi-retailer problem as follows.

[DOWMR]

$$\text{Minimize} \quad \sum_{t=1}^{T} \left(f_{0t} y_{0t} + \sum_{j \in J} \left\{ f_j y_{jt} + \sum_{r,s: r \leq s \leq t} H^j_{rst} x^j_{rst} \right\} \right) \tag{7.27}$$

$$\text{Subject to:} \quad \sum_{r,s: r \leq s \leq t} x^j_{rst} = 1, \quad j \in J, t = 1, \ldots, T, \tag{7.28}$$

$$\sum_{r: r \leq s} x^j_{rst} \leq y_{js}, \quad j \in J, t = 1, \ldots, T, s = 1, \ldots, t, \tag{7.29}$$

$$\sum_{s: r \leq s \leq t} x^j_{rst} \leq y_{0r}, \quad j \in J, t = 1, \ldots, T, r = 1, \ldots, t, \tag{7.30}$$

$$x^j_{rst}, y_{jt} \in \{0, 1\}, \; j \in J \cup \{0\}, s = 1, \ldots, T, \; r = 1, \ldots, s, t = s, \ldots, T. \tag{7.31}$$

Constraint set (7.28) ensures that each retailer's demand in every period is met by a pair of order periods (one from the warehouse and one from the retailer). Constraint sets (7.29) and (7.30) force the binary order variables in a period to one at the retailer and warehouse levels, respectively, if any order occurs in the associated period (as indicated by positive values of the appropriate x^j_{rst} variables). Arkin, Joneja, and Roundy [8] showed that the DOWMR problem is $\mathcal{NP}$-Complete.

Levi et al. [77] make mild assumptions on the structure of the h^j_{rst} values, which are satisfied by standard nonnegative linear holding costs in each period. These properties include nonnegativity, monotonicity in s and r, and the so-called Monge property. Monotonicity in s implies that for each retailer j, given a warehouse order period r and a demand period t, h^j_{rst} is either nondecreasing in s or nonincreasing in s for $r \leq s \leq t$. If a retailer exists such that this cost is nondecreasing in s, then under our standard linear holding cost assumptions, it is cheaper to hold inventory at the retailer than at the warehouse, all else being equal. Let J_R denote the set of such retailers. If this cost is nonincreasing in s, then it is cheaper to hold inventory at the warehouse than at the retailer, all else being equal. Let J_W denote the set of such retailers. Because of the stationary fixed order cost at each retailer, an optimal solution exists such that inventory for retailers in the set J_R is never held at the warehouse (any such inventory is transported immediately to the retailer after receipt at the warehouse). Note that for any retailer in the set J_R, given a solution in which the retailer orders but the warehouse does not, a solution with equal or lower cost exists such that the retailer's order is backed up to occur at the most recent warehouse order period. Thus, for such retailers, we only need x^j_{st} variables and h^j_{st} parameters.

Monotonicity in r implies that for any retailer order period s and demand period t, h^j_{rst} is nonincreasing in r for $r \leq s$, i.e., it costs more to hold inventory for a longer time. For retailers in the set J_R, because an optimal solution exists such that $r = s$ for such retailers (i.e., inventory is received at the warehouse and the retailer in the same period), if $r < r'$, then $h^j_{rrt} > h^j_{r'r't}$ (again, it costs more to hold inventory longer). The Monge property says that given a retailer $j \in J_W$ and a demand period t, then for $r_2 < r_1 \leq s_2 < s_1 \leq t$, we must have $h^j_{r_2 s_1 t} + h^j_{r_1 s_2 t} \geq h^j_{r_2 s_2 t} + h^j_{r_1 s_1 t}$, which implies that inventory at the warehouse serves retailers in first-in, first-out (FIFO) order.

Levi et al. [77] show that an optimal solution exists for the LP relaxation of DOWMR satisfying two key properties. The first of these is the Monge property of solutions, which implies that if $x^j_{rst} > 0$, and if $r' < r$ and $s' > s$, then $x^j_{r's't} = 0$. The second property is called the greedy property, which says that an optimal solution exists for the LP relaxation such that, given a retailer-period pair (j, t), demand is fulfilled in period t from more recent orders first. In other words, letting $(\hat{x}, \hat{y})$ denote an optimal solution to the LP relaxation, if the most recent retailer j order prior to or including period t occurs in period $s \leq t$ with $y_{js} = \hat{y}_{js} > 0$, then an optimal solution exists such that $\sum_{r=1}^{s} \hat{x}^j_{rst} = \hat{y}_{js}$. Thus, given a vector $\hat{y}^j$ of retailer order periods, we can allocate period t demand to retailer orders by starting with period t and going backwards, allocating a percentage equal to $\hat{y}_{js}$ of the demand in period t to the order in period s, until we have reached 100%. This implies that at most one period exists in a solution satisfying the greedy property with $0 < \sum_{r=1}^{s} \hat{x}^j_{rst} < \hat{y}_{js}$, and this period must be the first period to fractionally serve the demand in period t at retailer j. This also implies that if period s' is the earliest period to fractionally satisfy demand at retailer j in period t, then for each $s > s'$ such that $\hat{y}_{js} > 0$, we have $\sum_{r=1}^{s} \hat{x}^j_{rst} = \hat{y}_{js}$.

Given a solution $(\hat{x}, \hat{y})$ for the LP relaxation, Levi et al. [77] apply a rounding procedure, which we next summarize. The procedure begins by rounding the values of the (binary) warehouse order variables $\hat{y}_{0t}$, $t = 1, \ldots, T$. We start with an interval from 0 to $\sum_{r=1}^{T} \hat{y}_{0r}$, and associate the subinterval

$(\sum_{r=1}^{m-1} \hat{y}_{0r}, \sum_{r=1}^{m} \hat{y}_{0r}]$ with period m (note that this subinterval may have length zero for some values of m). Let $c \in (0,1]$ be a step parameter, let $W = \lceil (1/c) \sum_{r=1}^{T} \hat{y}_{0r} \rceil$, and note that the interval $(0, \sum_{r=1}^{T} \hat{y}_{0r}]$ is contained in the interval $(0, cW]$. Suppose we divide the interval $(0, cW]$ into contiguous subintervals of length c, with the lower limits of the intervals occurring at $0, c, 2c, \ldots, (W-1)c$. Next, shift the lower limit of each subinterval by a random value $\alpha_0 \in (0, c]$. The lower subinterval limits, called shift points, then become $\alpha_0, \alpha_0 + c, \ldots, \alpha_0 + (W-1)c$. If the interval $(\sum_{r=1}^{m-1} \hat{y}_{0r}, \sum_{r=1}^{m} \hat{y}_{0r}]$ contains any shift points, we then set $y_{0m} = 1$. Levi et al. [77] show that the expected warehouse ordering cost obtained via this randomized procedure is no more than $(1/c) \sum_{r=1}^{T} f_{0r} \hat{y}_{0r}$, i.e., no more than $1/c$ times the warehouse order costs in the optimal LP relaxation solution (when $c = 1$, this implies that the expected warehouse order costs in the randomized solution are no greater than those in the LP relaxation solution).

After determining the set of warehouse orders, we then move to determining the retailer orders for each $j \in J$. Given a set of warehouse order periods, the problem of determining an optimal set of order periods for each retailer decomposes by stage, and solving the problem for a given stage is equivalent to an instance of the UELSP (see Chapter 5), which can be solved efficiently via a shortest path algorithm. Guaranteeing a constant-factor performance bound, however, can evidently be achieved via a constructive randomized algorithm. Levi et al. [77] provide two such constructive randomized algorithms. The first of these uses the approach applied at the warehouse level with $c = 1$ to determine a set of candidate order periods for each $j \in J$. That is, $W_j = \lceil \sum_{s=1}^{T} \hat{y}_{js} \rceil$, and a shift point occurs at $\alpha_j + w$ for α_j selected uniformly from $(0, 1]$ and $w = 0, 1, \ldots, W_j - 1$.

If the interval $(\sum_{s=1}^{m-1} \hat{y}_{js}, \sum_{s=1}^{m} \hat{y}_{js}]$ contains any shift points, then period m is a *candidate* order period for retailer j. Next, suppose that for retailer j, period m is a candidate order period. If $j \in J_R$, then this retailer should only place an order when the warehouse places an order. If period m is a warehouse order period, then we also make it a permanent order period for retailer j; otherwise, we create two permanent order periods for retailer j: the latest warehouse setup period before period m and the earliest warehouse setup period after period m (if one exists). If $j \in J_W$, then we make period m a permanent order period for retailer j, along with the earliest warehouse setup period after period m, if one exists. Levi et al. [77] show that the expected retailer ordering costs associated with the permanent retailer order periods are at most twice those of the optimal LP relaxation solution. Next, they show that the expected holding costs associated with this solution approach are bounded by twice the holding costs incurred in the optimal LP relaxation solution. This implies that this randomized solution algorithm has an expected solution value that is no more than twice the optimal solution value.

When $c = 1$, the expected warehouse order costs are no more than those incurred in the optimal LP relaxation, while the expected retailer order and holding costs are no more than twice those in the LP relaxation. Because of this imbalance, Levi et al. [77] provide a rounding approach using $c \in (0, 1/2]$. The approach to setting warehouse order periods is the same as in the previous approach, while the approach for setting retailer order periods differs slightly. Using a step parameter equal to $1 - c$, suppose period m is a candidate order period for the retailer using the previously described approach. Then if $j \in J_R$, a permanent order period for the retailer is created in the earliest warehouse order period after m. For $j \in J_W$, period m itself becomes a permanent order period. The expected retailer order costs under this approach are bounded by $1/(1-c)$ times those incurred in the optimal LP relaxation solution. Levi et al. [77] also show that the expected holding costs incurred via this modified procedure are no more than $1/(1-c)$ times those incurred in the optimal LP relaxation solution. As a result, if $c = 1/2$, this approach also gives a worst-case bound of 2 on the expected cost of the solution.

Suppose we use $c = 1/3$ under the latter approach. Then the expected cost of this approach is no more than three times the fixed warehouse order costs of the LP relaxation solution, denoted by F^W_{LP}, plus 3/2 times the sum of retailer order and holding costs of the LP relaxation solution, denoted by $F^R_{LP} + H^R_{LP}$, and we write the resulting total expected cost bound as $3F^W_{LP} + (3/2)(F^R_{LP} + H^R_{LP})$. Using the first approach with $c = 1$ gives a bound on the expected solution value of $F^W_{LP} + 2(F^R_{LP} + H^R_{LP})$. Suppose we apply both approaches and take the minimum cost between the two. Because $\min\{a, b\} \leq \lambda a + (1-\lambda)b$ when $0 \leq \lambda \leq 1$, using $\lambda = 3/5$ gives a minimum less than or equal to $(9/5)(F^W_{LP} + F^R_{LP} + H^R_{LP})$.[3] Thus, the strategy of applying both approaches and taking the minimum gives an expected cost of no more than 1.8 times the optimal solution value.

Next, we derandomize the algorithm. Note that given the warehouse orders, we can solve each retailer's problem optimally for the fixed set of warehouse orders in polynomial time, and the resulting cost will be no more than that of the randomized algorithm. Because of this, we only need to derandomize the approach used to obtain warehouse order periods. Recall that this approach worked by considering whether the interval $(\sum_{r=1}^{m-1} \hat{y}_{0r}, \sum_{r=1}^{m} \hat{y}_{0r}]$ for $m = 1, \ldots, T$ contains any shift points; if so, we then set $y_{0m} = 1$. The shift points are given by $\alpha_0, \alpha_0 + c, \ldots, \alpha_0 + (W-1)c$ for some $c \in (0, 1]$ and α_0 uniformly selected from $(0, c]$, and the maximum possible number of shift points is no more than $\lceil T/c \rceil$. Increasing α_0 from 0 to c is equivalent to moving each shift point to the right a distance equal to c. As we do this, the resulting vector of warehouse orders may change when a shift point crosses a boundary of an interval of the form $(\sum_{r=1}^{m-1} \hat{y}_{0r}, \sum_{r=1}^{m} \hat{y}_{0r}]$.

---

[3] We would like to provide an upper bound on the minimum between $F^W_{LP} + 2(F^R_{LP} + H^R_{LP})$ and $3F^W_{LP} + (3/2)(F^R_{LP} + H^R_{LP})$. Letting $c_W = F^W_{LP}$ and $c_R = F^R_{LP} + H^R_{LP}$, this is equivalent to providing a bound on $\min\{c_W + 2c_R; 3c_W + (3/2)c_R\}$. Using $\min\{a, b\} \leq \lambda a + (1-\lambda)b$ when $0 \leq \lambda \leq 1$, the quantity $\lambda(c_W + 2c_R) + (1-\lambda)(3c_W + (3/2)c_R) = (3 - 2\lambda)c_W + (3/2 + \lambda/2)c_R$ provides such a bound for any $0 \leq \lambda \leq 1$. Observe that $(3 - 2\lambda)c_W + (3/2 + \lambda/2)c_R \leq \max\{3 - 2\lambda; 3/2 + \lambda/2\}(c_W + c_R)$. It is straightforward to show that the minimum value of $\max\{3 - 2\lambda; 3/2 + \lambda/2\}$ such that $0 \leq \lambda \leq 1$ occurs when $3 - 2\lambda = 3/2 + \lambda/2$, i.e., at $\lambda = 3/5$. Thus, $(9/5)(c_W + c_R) = (9/5)(F^W_{LP} + F^R_{LP} + H^R_{LP})$ provides the desired bound.

Note that at most one shift point may cross a given boundary point as we increase α_0 from 0 to c. In the worst case, as we increase α_0 from 0 to c, a shift point may start below the lower limit of a given interval that contains no shift points, may then increase to a value between the interval's lower and upper limits, and then subsequently increase to a value above the interval's upper limit; alternatively, a shift point may begin with a value between an interval's lower and upper limits, may increase to a value above the interval's upper limit, and the next lower shift point may subsequently increase to a value above the interval's lower limit (note that each of these changes corresponds to α_0 crossing an interval boundary). In the former case, the order associated with the interval begins as "off," is subsequently turned "on," and is later turned off again; in the latter case, the order associated with the interval begins as on, is turned off, and later turned on again. Thus, as we increase α_0 from 0 to c, a given setup may, in the worst case, go through the states off-on-off or on-off-on. The worst-case number of order vectors we need to consider at the warehouse corresponds to the case in which, as we increase α_0 from 0 to c, each order changes state (from off-to-on or from on-to-off) at most twice. Thus, the worst case is no worse than the case in which all orders begin in the off state, each is turned on at a unique value of α_0 and is later turned off at some value of α_0; this leads to no more than $2T$ order vectors to consider. Because each of these changes in order vectors occurs only when α_0 crosses some interval boundary, it is sufficient to consider the value of α_0 at each interval boundary. The value of α_0 at the upper limit of the interval $(\sum_{r=1}^{m-1} \hat{y}_{0r}, \sum_{r=1}^{m} \hat{y}_{0r}]$ equals $(\sum_{r=1}^{m} \hat{y}_{0r}) \pmod{c}$, and we can compute the corresponding "critical" value of α_0 for each interval in order to capture each of the unique warehouse order vectors that results from the randomized algorithm. The choice of α_0 that gives the minimum cost solution will produce a solution with a cost that is no more than the expected value across all uniformly selected α_0 values. This implies that a deterministic polynomial time algorithm exists for the DOWMR problem that is guaranteed to give a solution with value no more than 1.8 times the optimal solution value.

Example 7.1 To illustrate the rounding method for the DOWMR problem, consider a six-period example with two identical retailers. The problem parameters are shown in the table below, where $h_t^{j\prime}$ denotes the local holding cost at stage j, for $j = 0, 1, 2$. Assuming that $d_{jt} = 1$ in each period and for $j = 1, 2$, then $H_{rst}^j = \sum_{\tau=r}^{s-1} h_\tau^{0\prime} + \sum_{\tau=s}^{t-1} h_\tau^{j\prime}$ for $j = 1, 2$. In this example, the retailers' fixed order costs equal $f_1 = f_2 = 4$ in each period.

Period, t	1	2	3	4	5	6
f_{0t}	0	5	10	12	6	3
$h_t^{j\prime}$, $j = 1, 2$	6	5	4	3	2	0
$h_t^{0\prime}$	3	2	2	1	1	0

The LP relaxation of this instance of the DOWMR problem has an optimal solution value of 65.5, with $\hat{y}_0 = (1, 1, 0, 0.5, 0.5, 0)$, $\hat{y}_1 = (1, 1, 0.5, 0.5, 0.5, 0.5)$, and $\hat{y}_2 = (1, 1, 0.5, 0.5, 1, 0.5)$. We focus on the rounding procedure as applied to the warehouse, as we can optimize the retailers' costs in polynomial time for any given binary solution in the y_{0t} variables. The vector $\hat{y}_0$ leads to associating the interval $(0, 1]$ with period 1, the interval $(1, 2]$ with period 2, the interval $(2, 2.5]$ with period 4, and the interval $(2.5, 1]$ with period 5. Note that the intervals associated with periods 3 and 6 are empty.

Suppose we use $c = 1/3$ and $\alpha_0 = \epsilon$, where $\epsilon > 0$ and arbitrarily small. We set $W = \lceil (1/c) \sum_{t=1}^{T} y_{0t} \rceil = 9$, and the resulting shift points occur at $\epsilon, 1/3 + \epsilon, 2/3 + \epsilon, 1 + \epsilon, 4/3 + \epsilon, 5/3 + \epsilon, 2 + \epsilon$, $7/3 + \epsilon$, and $8/3 + \epsilon$, which implies that shift points fall on the intervals associated with periods 1, 2, 4, and 5, leading to the warehouse order vector $y_0 = (1, 1, 0, 1, 1, 0)$. Optimizing the retailer's order sequence with this warehouse order vector gives a solution with objective function value 70. Note that any $\alpha_0 \in (0, 1/3]$ will produce the same warehouse order vector with $c = 1/3$.

Suppose instead that $c = 1$ and $\alpha_0 = \epsilon$. We set $W = 3$, and the resulting shift points occur at $\epsilon, 1 + \epsilon$, and $2 + \epsilon$, which implies that shift points fall on the intervals associated with periods 1, 2, and 4, leading to the warehouse order vector $y_0 = (1, 1, 0, 1, 0, 0)$. Optimizing the retailer's order sequence with this warehouse order vector gives a solution with objective function value 67. If we set $\alpha_0 = 0.5 + \epsilon$, then the shift points occur at $0.5 + \epsilon, 1.5 + \epsilon, 2.5 + \epsilon$, which implies that shift points fall on the intervals associated with periods 1, 2, and 5, leading to the vector $y_0 = (1, 1, 0, 0, 1, 0)$. Optimizing the retailer's order sequence with this warehouse order vector also gives a solution with objective function value 66, which happens to equal the optimal solution for this instance of the DOWMR problem. □

We next briefly discuss the dynamic joint replenishment problem (DJRP), which is a multiple-item, single-stage inventory planning problem in which each item incurs fixed order costs as well as inventory holding costs. In addition to item-specific fixed order costs, an additional "major" fixed order cost is incurred for placing any order, which is independent of the mix and quantity of items ordered. In the absence of this additional major fixed order cost, we can solve the UELSP for each item separately. However, the inclusion of the major fixed order cost makes the problem considerably more difficult (Arkin et al. [8] show that this problem is $\mathcal{NP}$-Complete). We can model the DJRP as a special case of the DOWMR problem by assuming that the warehouse incurs the major fixed order cost, and that the holding cost at the warehouse is infinite in each period. If this is the case, then an order will have to be placed at the warehouse any time a downstream stage places an order. Thus, as a corollary to the result Levi et al. [77] obtained for the DOWMR problem, we also have a constant-factor approximation algorithm with a performance bound of 1.8 for the DJRP.

A stream of literature considers the DOWMR problem when the warehouse holds no inventory and serves strictly as a cross-docking facility. Chan et al.

[22] considered this problem class under piecewise-linear (but nonlinear overall) costs. They showed that the optimal ZIO solution has a value of no more than 4/3 the optimal solution value (this bound decreases to 5.6/4.6 if the warehouse order costs are time-invariant). Shen, Shu, Simchi-Levi, Teo, and Zhang [105] later considered this problem class with general piecewise-linear cost structures and provided a solution method that guarantees a solution value of $(1+\epsilon)$ times the optimal solution (with $\epsilon > 0$) in polynomial time in the number of retailers, time periods, and $1/\epsilon$ (therefore, the associated algorithm is an FPTAS).

We note that Levi, Roundy, and Shmoys [76] developed a primal-dual algorithmic approach for the DJRP and assembly problems with non-speculative costs that gives a constant-factor approximation bound of 2 for these problem classes. In addition, Roundy [101] considered dynamic planning problems with general network structures under stationary costs and time-varying demands (this version of the problem also allowed for joint replenishment costs within product families). He developed heuristic solutions satisfying the nestedness and ZIO properties, and requiring that certain subsets of customers order together, in clusters. Roundy [101] provides a heuristic method that produces a solution with a value no more than twice that of an optimal nested policy. Finally, a sizable literature exists that considers dynamic multistage production and distribution planning problems with finite stage capacities, although scope limitations do not permit covering this class of problems within this book. We refer the reader to the excellent survey paper on this stream of work by Buschkühl, Sahling, Helber, and Tempelmeier [20].

7.3 Models with Constant Demand Rates

This section discusses multistage production and inventory planning problems in which the demand arising at any stage occurs at a constant rate, and all problem cost parameters are time invariant as well. We begin by assuming basic familiarity with the single-stage EOQ model (Harris [65]), a continuous-time model with a constant demand rate of λ units per unit time for a single item, a fixed charge of f for a procurement order (or production setup), a variable cost of c per unit ordered (or produced), and a holding cost of h' per unit held per unit time. For convenience we will assume a base time unit of a year, although any base time unit may be used in defining λ and h', provided that the same base unit is used for both. The EOQ model assumes the system operates over an infinite horizon, a constant production or order lead time exists, all cost and demand parameters are constant, and no shortages are allowed. For the EOQ model, an optimal solution exists with a time-invariant lot size equal to Q and such that when the inventory on hand hits zero, a new batch of size Q arrives immediately. The constant demand rate and equal batch sizes lead to a sawtooth inventory curve (see Figure 7.3) that has a corresponding average annual cost equation equal to

$$c\lambda + \frac{f\lambda}{Q} + \frac{h'Q}{2} = c\lambda + \frac{f}{T} + g'T, \tag{7.32}$$

where $T = Q/\lambda$ is the time between batch order arrivals when using a batch size of Q and $g' = h'\lambda/2$. This cost equation is convex in T for $T \geq 0$ (or in Q for $Q \geq 0$) with a global minimum value at

$$T^* = \sqrt{\frac{f}{g'}}, \tag{7.33}$$

with $Q^* = \sqrt{f\lambda^2/g'} = \sqrt{2f\lambda/h'}$.

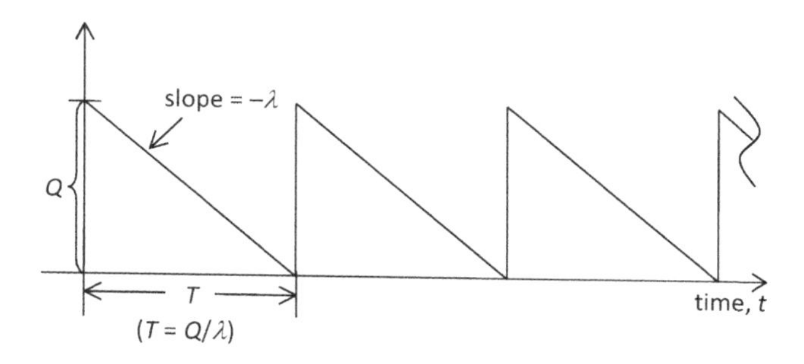

FIGURE 7.3
Sawtooth curve for inventory level in the EOQ Model.

In the single-stage EOQ problem, the demand at the single stage occurs at a constant rate. Suppose we have a two-stage serial system in which constant external demand occurs at a rate λ at stage 2, while stage 2 replenishes its inventory in batches of size Q, which are supplied every T time units from stage 1. Consider the inventory level over time at stage 1, which replenishes its stock using an external supplier with a fixed lead time and unlimited capacity. It is optimal for the batch size at stage 1 to take a value equal to an integer multiple of Q,[4] where the use of a constant batch size at each stage is known as a *stationary* policy (i.e., the time between orders is a fixed value for all orders at all stages).[5] Observe that if this multiple equals $M > 1$, then inventory at stage 1 falls by Q units every $T = Q/\lambda$ time units and remains flat in

[4]If the batch size at stage 1 is not an integer multiple of Q, then this violates the zero-inventory ordering (ZIO) property, and it is straightforward to show that a solution exists satisfying the ZIO property with lower average cost per unit time.

[5]Although stationary policies are not guaranteed to be optimal for all network structures, they are practical and convenient in implementation, and are optimal for serial systems with constant demand rates and cost parameters (see Zipkin [120]). Our analysis in this section will therefore consider only policies that are stationary at each stage.

between shipments to stage 2. The sum of the inventory levels at stages 1 and 2, however, still falls at a constant rate λ.

By using the concepts of echelon inventory and echelon holding cost (defined in Section 7.2.2), we arrive at a more convenient way of expressing the total system cost. In our two-stage system, the echelon holding cost for stage 1 equals the installation (local) holding cost (the traditional definition of holding cost we have used), while the echelon holding cost at stage 2 equals the installation holding cost at stage 2 less the echelon holding cost at stage 1. Observe that the graph of echelon inventory level at any stage follows a saw-tooth pattern as in the single-stage EOQ model. Figure 7.4 shows the local and echelon inventory levels for stage 1, which serves stage 2 in a two-stage serial system with a demand rate of λ, and where the order quantity at stage 1, Q_1, equals twice the order quantity at stage 2, Q_2.

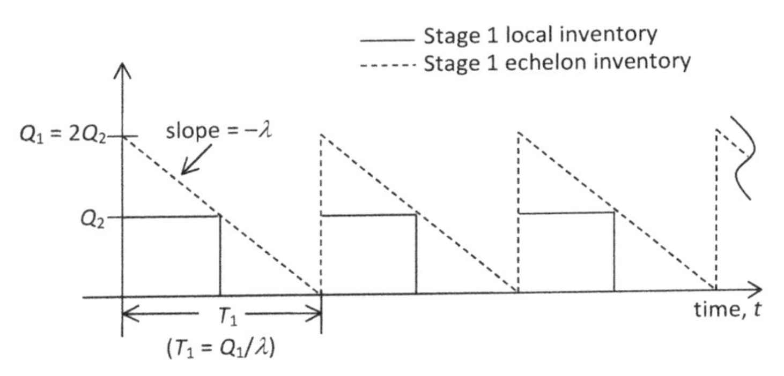

FIGURE 7.4
Echelon and local inventory pattern for stage 1 in a two-stage serial system with $Q_1 = 2Q_2$.

7.3.1 Stationary, nested, power-of-two policies

For a multistage system network, let I denote the set of stages (and the corresponding nodes in the network), indexed by i and j, and let A denote the arc set, where arc $(i, j) \in A$ exists if stage j orders from stage i. Let h_i denote the echelon holding cost rate at stage i, with T_i equal to the time between ordering at stage i and λ_i equal to the rate of demand for stage i echelon inventory; as earlier, we let $g_i = h_i\lambda_i/2$.

For a serial or assembly system where demand occurs only at the final downstream stage, Crowston, Wagner, and Williams [35] showed that within the class of stationary policies, nested policies are optimal. Recall that a policy is nested if placing an order at a stage implies that all direct successors must also place an order at the same time, although a successor stage may order more frequently than a predecessor. This implies that in an assembly or serial system with stationary reorder intervals, if stage j is the (single) immediate successor of stage i, and stage j orders Q units every T time units, then stage i will order MQ units every MT time units, where M is an integer greater than or equal to one. Unfortunately, for general or distribution systems, nested policies may perform arbitrarily poorly (see Roundy [98]). To show the intuition behind this, Roundy [98] considers a system in which a warehouse serves downstream stages 1 and 2, where the warehouse and stage 1 have fixed order costs equal to 1 and stage 2 has a fixed order cost equal to K. Suppose that K is much larger than 1, the demand rate at stage 2 is much smaller than that at stage 1, and the holding costs are the same at stages 1 and 2. Then in an optimal solution, the warehouse and stage 1 will place replenishment orders much more frequently than stage 2. However, in a nested policy, stage 2 must place replenishment orders when the warehouse does; thus, forcing this requirement on the system may result in a severely suboptimal solution. Because stationary and nested policies are often found in practice due to their ease of implementation and effectiveness for serial and assembly systems, we will discuss their application to general systems. Later we will consider the implications of relaxing the nestedness requirement via a discussion of the approach that Roundy [98] provided for one-warehouse multi-retailer distribution systems.

Suppose that in addition to requiring a stationary, nested policy, we also require that the replenishment interval at each stage i, T_i, takes a value that is a multiple of some base planning interval length equal to T_L (e.g., one day, one week, etc.). For now, assume that T_L is a fixed value (we will later discuss optimizing the value of T_L). For stage i, let M_i denote this multiple, i.e., we require $T_i = M_i T_L$ with $M_i \in \mathbb{Z}_+$ for each $i \in I$, where $\mathbb{Z}_+$ denotes the set of nonnegative integers. Ensuring that a solution satisfies the "nestedness" property requires that for each arc (i, j), M_i is evenly divisible by M_j. Rather than attempt to explicitly express the set of all solutions that satisfy this requirement, Maxwell and Muckstadt [85] impose an additional requirement that both enables expressing the problem in a compact form and leads to surprising results on the quality of solutions obtained. This requirement forces each M_i to take a value equal to a *power-of-two* multiple of the base planning period length T_L, i.e., $M_i = 2^{k_i} T_L$ with $k_i \in \mathbb{Z}_+$ for each $i \in I$. This, along with the requirement that $M_i \geq M_j$ for each $(i, j) \in A$ ensures that the resulting solution will be nested. At first blush, this power-of-two requirement may appear unnecessarily restrictive. However, as we later discuss, the deviation of such a policy from optimality among all nested policies turns out to be surprisingly small. It should be noted that the seminal results in Roundy [98] and Maxwell and Muckstadt [85] were developed independently at around the same time. Both of these works demonstrated the relative effectiveness of power-of-two policies for multistage inventory problems, which has led to numerous subsequent works that analyze and demonstrate the effectiveness and benefits of this class of policies in various settings.

The resulting multistage EOQ (MSEOQ) problem formulation may be expressed as follows:

$$\textbf{[MSEOQ]} \quad \text{Minimize} \quad \sum_{i \in I} \left\{ \frac{f_i}{T_i} + g_i T_i \right\} \tag{7.34}$$

$$\text{Subject to:} \quad T_i = 2^{k_i} T_L, \qquad i \in I, \tag{7.35}$$

$$k_i \geq k_j, \qquad (i,j) \in A, \tag{7.36}$$

$$k_i \in \mathbb{Z}_+, \qquad i \in I. \tag{7.37}$$

The MSEOQ problem is a nonlinear integer optimization problem with a separable convex objective function in the T_i variables. Maxwell and Muckstadt [85] begin by solving a relaxation of the problem, and then show that the solution to this relaxed problem can be used as a starting point for efficiently solving the MSEOQ problem. The relaxed problem eliminates the requirement that each stage's reorder interval must take a value equal to an integer multiple of the base planning period length T_L and requires only that each stage's reorder interval take a value at least as great as that for all immediate successors. The relaxed problem, RP, is formulated as follows:

$$\textbf{[RP]} \quad \text{Minimize} \quad \sum_{i \in I} \left\{ \frac{f_i}{T_i} + g_i T_i \right\} \tag{7.38}$$

$$\text{Subject to:} \quad T_i \geq T_j \geq 0, \qquad (i,j) \in A. \tag{7.39}$$

The relaxed problem RP is a convex optimization problem with a linear feasible region, which implies that the Karush Kuhn-Tucker (KKT) optimality conditions[6] are necessary and sufficient (see Bazaraa et al. [15]). To characterize the KKT conditions, we require defining a nonnegative KKT multiplier θ_{ij} for each constraint of the form $T_i \geq T_j$, $(i,j) \in A$. In order to keep track of the KKT multipliers associated with a node $i \in I$ in the network, it is convenient to define the set of direct predecessors of i, denoted by $\mathcal{P}(i)$, and the set of direct successors of i, denoted by $\mathcal{S}(i)$. In addition to the feasibility constraints (7.39) and the nonnegativity of the KKT multipliers, the KKT conditions require

$$\frac{f_i}{T_i^2} - g_i + \sum_{j \in \mathcal{S}(i)} \theta_{ij} - \sum_{j \in \mathcal{P}(i)} \theta_{ji} = 0, \quad i \in I, \tag{7.40}$$

$$\theta_{ij}(T_i - T_j) = 0, \qquad (i,j) \in A, \tag{7.41}$$

[6]For the problem $\min f(x) : g_i(x) \leq 0, i = 1, \ldots, m$, where $f(x)$ is a function of the n-vector x in $\mathbb{R}^n$ and $g_i(x)$ denotes the ith constraint function, the KKT conditions are expressed as follows. Let u denote an m-vector of KKT multipliers associated with the m constraints $g_i(x) \leq 0$, $i = 1, \ldots, m$. Let $\nabla f(x)$ ($\nabla g(x)$) denote the gradient vector (matrix) of $f(x)$ ($g(x)$) at x. The KKT conditions are written as $g_i(x) \leq 0$, $i = 1, \ldots, m$ (analogous to primal feasibility), $u_i g_i(x) = 0$, $i = 1, \ldots, m$ (analogous to complementary slackness), and $\nabla f(x) + u \nabla g(x) = 0$ with $u \geq 0$ (analogous to dual feasibility). When $f(x)$ is a convex function and each $g_i(x)$, $i = 1, \ldots, m$, is a convex function, the KKT conditions are sufficient for optimality. If, in addition, the feasible region contains an interior point or corresponds to the intersection of a finite set of linear constraints (i.e., consists of a polyhedron), the KKT conditions are also necessary for optimality (these *constraint qualifications* that lead to the necessity of the KKT conditions may be expressed in various forms; see [15]).

where (7.40) is equivalent to

$$T_i = \sqrt{\frac{f_i}{g_i - \sum_{j \in \mathcal{S}(i)} \theta_{ij} + \sum_{j \in \mathcal{P}(i)} \theta_{ji}}}. \tag{7.42}$$

Note the similarity between (7.42) and the single-stage EOQ formula (7.33). If stage i does not share the same reorder interval with any of its predecessors or successors, implying $\theta_{ij} = 0$ for all $j \in \mathcal{P}(i) \cup \mathcal{S}(i)$ as a result of (7.41), then T_i equals the single-stage EOQ value at stage i. The KKT multipliers therefore serve to adjust the single-stage EOQ value when a stage shares a reorder interval with a set of predecessors and/or successors that do not have equal values of f_i/g_i (note that if a stage i and a successor stage j exist with $f_i/g_i = f_j/g_j$, then it is possible to have $T_i = T_j$ *and* $\theta_{ij} = 0$). That is, the KKT multipliers θ_{ij} serve to *equalize* the reorder intervals for stages that share the same reorder interval but differ in their ratios of f_i to g_i.

The approach Maxwell and Muckstadt [85] use for solving RP begins with a feasible solution in which all T_i values are equal, although this solution may violate the KKT conditions. This solution is obtained by summing (7.40) over all $i \in I$, and letting $T_i = \hat{T}$ for all $i \in I$. Because all of the θ_{ij} terms sum to zero, the resulting solution is given by

$$\hat{T} = \sqrt{\frac{\sum_{i \in I} f_i}{\sum_{i \in I} g_i}}. \tag{7.43}$$

If a set of nonnegative θ_{ij} values exists that satisfies (7.40) when $T_i = \hat{T}$ for all $i \in I$, then we have an optimal solution to RP. If this is not the case, then we cannot have equal reorder intervals at all stages in an optimal solution. In this case, we must divide the network into disjoint subnetworks, where each stage is in one subnetwork, all of the stages within a subnetwork share a common reorder interval, and we maintain $T_i \geq T_j$ for all $(i, j) \in A$. If we can do this in a way that maintains feasibility for RP and improves upon the existing solution, we will have moved to a better feasible solution.

Consider a division of the stages into two disjoint subsets I^+ and I^-, such that $I^+ \cup I^- = I$, where the only arcs in the network connecting these subsets are directed from nodes in I^+ to nodes in I^-. If we increase T_i for $i \in I^+$ from the current value of $\hat{T}$ and decrease T_i for $i \in I^-$ from the current value of $\hat{T}$, we maintain feasibility for RP. Note that the partial derivative of the objective of RP with respect to T_i equals $-f_i/T_i^2 + g_i$. Therefore, at the current solution, the marginal decrease in the objective of RP if we *increase* each T_i for $i \in I^+$ from the current value of $\hat{T}$, which we denote by $\Delta(I^+, I^-)$, is given by

$$\Delta(I^+, I^-) = \sum_{i \in I^+} \left\{ \frac{f_i}{\hat{T}^2} - g_i \right\}. \tag{7.44}$$

Similarly, the marginal decrease in the objective of RP if we *decrease* each T_i

for $i \in I^-$ from the current value of $\hat{T}$ equals

$$\sum_{i \in I^-} \left\{ -\frac{f_i}{\hat{T}^2} + g_i \right\}, \tag{7.45}$$

which also equals $\Delta(I^+, I^-)$, because $\sum_{i \in I} \{f_i/\hat{T}^2 - g_i\} = 0$. If a division of I into subsets I^+ and I^- exists with a positive $\Delta(I^+, I^-)$, then an improving feasible solution exists in which we increase the reorder interval for $i \in I^+$ and decrease it for $i \in I^-$. As a result, we would like to determine the division, or cut, of I^+ and I^- with the maximum value of $\Delta(I^+, I^-)$. Maxwell and Muckstadt [85] describe the solution of this maximal cut problem by solving a minimal flow problem on a graph with lower bounds on flow. We will next describe this problem in a slightly different way, using a minimum cost network flow approach that leads to equivalent results.

Consider the form of (7.40), which looks suspiciously like a network flow balance constraint at each node, and where the θ_{ij} values correspond to arc flow variables. For fixed values of the T_i variables for all $i \in I$ (e.g., $T_i = \hat{T}$ for all $i \in I$) we can view (7.40) as a set of flow balance constraints for a minimum cost network flow problem in which node i has net supply equal to $g_i - f_i/\hat{T}^2$ if this value is positive and net demand equal to $f_i/\hat{T}^2 - g_i$ if this value is positive. Letting, $a_i = f_i/\hat{T}^2 - g_i$, this is equivalent to saying that a node has a net supply of $-a_i$ if $a_i < 0$ (we will refer to such nodes as net-supply nodes) and a net demand of a_i if $a_i > 0$ (we will refer to such nodes as net-demand nodes). Let I_S (I_D) denote the set of net-supply (net-demand) nodes. We rewrite (7.40) as

$$\textstyle\sum_{j \in \mathcal{P}(i)} \theta_{ji} - \sum_{j \in \mathcal{S}(i)} \theta_{ij} = a_i, \quad i \in I, \tag{7.46}$$

which looks like a standard set of network flow balance constraints. Again, if we can find nonnegative θ_{ij} flow variables that satisfy these constraints and such that the a_i values imply $T_i \geq T_j$ for all $(i,j) \in A$, then we will have a KKT solution. Unfortunately, we are not guaranteed that the a_i values we are using admit a set of feasible nonnegative flows in the network. Because of this, we create a modified network by adding two nodes to the network, one source node (node 0) and one sink node (node $|I|+1$). We add $|I|+1$ arcs from the source to every other node in the network, and $|I|$ additional arcs from each node in the original network to the sink node. For each $i \in I_D$, let a_i equal the node's demand, and for each $i \in I_S$, let $-a_i$ equal the node's supply. The source node has a supply equal to $\sum_{i \in I_D} a_i$, which equals the sink node's demand. Suppose we minimize the sum of the flows on arcs from node zero to nodes 1 through $|I|$, subject to (7.46) and nonnegativity of the flow variables, which is equivalent to maximizing the flow on the arc from the source to sink node while obeying these constraints.

If a feasible solution exists with zero flow on arcs from node zero to nodes 1 through $|I|$ (and a flow of $\sum_{i \in I_D} a_i$ directly from the source to the sink), then we have identified a KKT point, i.e., a nonnegative set of feasible flows on the original network. Otherwise, the flow from the source is split between the sink node and original nodes in the network. Suppose this is the case. Then the arcs with positive flow in the network in an optimal solution to this network flow problem induce a spanning tree, and this spanning tree includes the arc from the source directly to the sink. If we were to cut the source-to-sink arc from this spanning tree, we would be left with two subtrees, one connected to the sink and one connected to the source. The subtree connected to the source has *excess demand*, which must be met using flow from the source, and we refer to this subtree as the excess-demand subnetwork. The subtree connected to the sink has *excess supply*, which must flow to the sink, and we refer to this subtree as the excess-supply subnetwork. Observe that we cannot have arcs directed in the original network from the excess-supply subnetwork to the excess-demand subnetwork; otherwise, this imbalance would be resolved without using supply from the source node. In other words, arcs in the original network must be directed from the excess-demand subnetwork to the excess-supply subnetwork.

How can we resolve these excesses? This is done by increasing T_i for all nodes in the excess-demand subnetwork, and decreasing T_i for all nodes in the excess-supply subnetwork. When we increase T_i for a net-demand node, this reduces the node's demand level. When we increase T_i for a net-supply node, we increase the node's supply level. Thus, by increasing T_i for all nodes in the excess-demand subnetwork, we simultaneously reduce the demand of each net-demand node and increase the supply of each net-supply node in the subnetwork, which reduces the excess demand of the subnetwork. Similarly, by decreasing T_i for all nodes in the excess-supply subnetwork, we reduce the supply of each excess-supply node and increase the demand at each excess-demand node. This permits reducing the degree of excess supply and excess demand in each subnetwork. Fortunately, the direction of change within each subnetwork is consistent with our requirement that $T_i \geq T_j$ for each $(i, j) \in A$, as arcs are directed from the excess-demand subnetwork to the excess-supply subnetwork in the original network. If we refer to nodes in the excess-demand subnetwork as the set I^+ and those in the excess-supply subnetwork as I^-, then the flow on the arcs from the source node to nodes in I^+ precisely equals $\Delta(I^+, I^-)$. To see this, observe that the flow to the nodes in I^+ from the source node in the network flow problem must equal the excess demand in the subnetwork, $\sum_{i \in I^+} a_i$, which is precisely Equation (7.44). The solution approach we have described for this network flow problem provides the same solution as the minimum flow problem solved by Maxwell and Muckstadt [85], and the separation of I^+ and I^- identified by this flow problem also corresponds to the maximal cut, with cut value $\Delta(I^+, I^-)$.

Once we have identified I^+ and I^-, we have determined that none of the nodes in I^+ will share a reorder interval with a node in I^-, i.e., $T_i > T_j$ for $i \in I^+$ and $j \in I^-$. This implies that the value of θ_{ij} associated with an arc (i, j) with $i \in I^+$ and $j \in I^-$ will equal zero. Thus, these two subnetworks are decoupled, i.e., we can analyze each subnetwork separately as its own network, eliminating the flow variables that connect nodes in I^+ to I^-. Because of this, within each subnetwork, as in the original network, if we sum KKT condition (7.40) over all nodes in the subnetwork, the θ_{ij} variables will sum to zero, and we arrive at order intervals equal to $\hat{T}^+ = \sqrt{\sum_{i\in I^+} f_i / \sum_{i\in I^+} g_i}$ and $\hat{T}^- = \sqrt{\sum_{i\in I^-} f_i / \sum_{i\in I^-} g_i}$. We then find the maximal cut for each subnetwork by solving the network flow problem for each subnetwork as described earlier. This process is repeatedly applied by continuing to subdivide networks until all subnetworks have nonpositive maximal cuts, at which point problem RP is solved. This is equivalent to the situation in which no flow exists from the source node to any original node in every subnetwork flow problem.

Example 7.2 We illustrate this procedure with an example based on the assembly structure shown in Figure 7.5. The numbers shown below the nodes in Figure 7.5 indicate the fixed order costs, and we assume that $g_i = 1$ for all $i = 1, \ldots, 5$. The initial solution for $\hat{T}$ equals 2.49. This value of $\hat{T}$ results in the network flow problem shown in Figure 7.6, where the values of a_i are shown for each node. A negative a_i indicates that the node is a net-supply node, while a positive a_i implies a net-demand node.

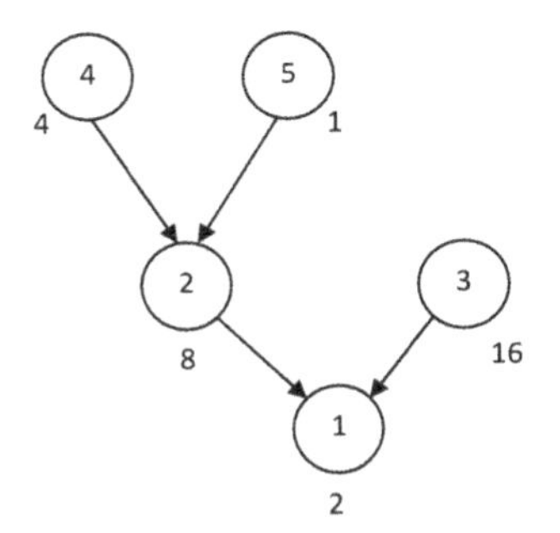

FIGURE 7.5
Five-stage assembly system.

The solution that minimizes the sum of the flows from the source to nodes 1 through 5 is shown in Figure 7.7. In this solution, node 3 is supplied by the source node, while nodes 1, 2, 4, and 5 are connected via a tree to the sink node. This implies that $I^+ = \{3\}$, while $I^- = \{1, 2, 4, 5\}$. We therefore subdivide the graph into I^+ and I^-. Since I^+ has only one element, it will not be divided further, and note that $T_3 = \hat{T}^+ = \sqrt{16} = 4$. We next consider whether I^- must be further subdivided. The network flow problem for this subgraph is shown in Figure 7.8, where $\hat{T}^- = 1.94$, and the newly computed a_i values are shown next to each node. The solution that minimizes the flow to nodes 1, 2, 4, and 5 from the source is shown in Figure 7.9. In this solution, nodes 2, 4, and 5 form a tree connected to the source node, while node 1 is connected to the sink node. We thus define $I^{-+} = \{2, 4, 5\}$ and $I^{--} = \{1\}$. The network $I^{--} = \{1\}$ cannot be further subdivided, and note that $T_1 = \hat{T}^{--} = \sqrt{2}$.

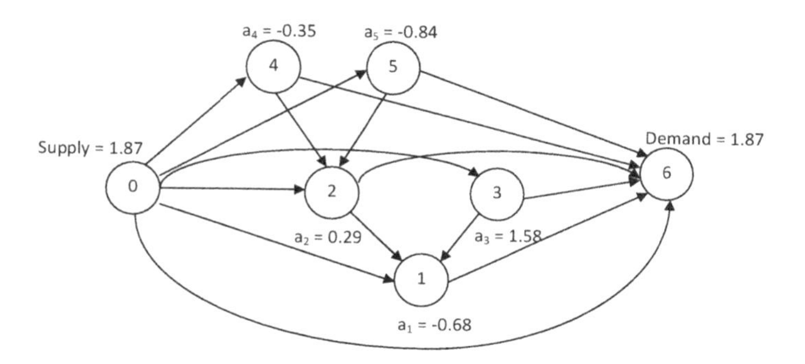

FIGURE 7.6
Network flow problem for assembly problem example.

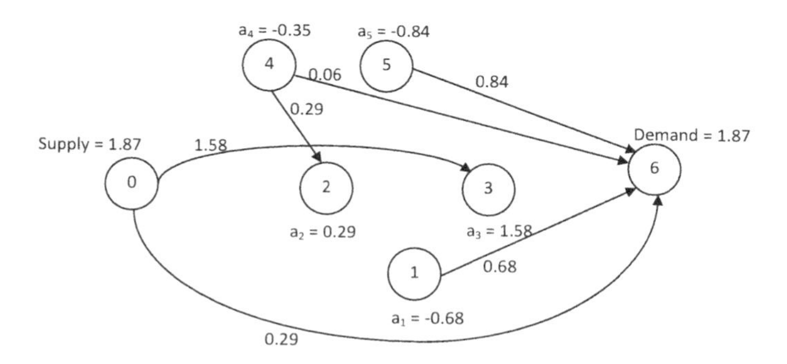

FIGURE 7.7
Network flow problem solution for flow problem shown in Figure 7.6.

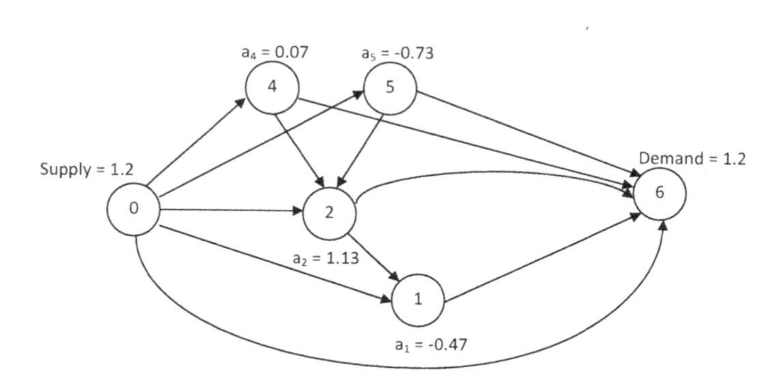

FIGURE 7.8
Network flow problem for I^- resulting from assembly problem example.

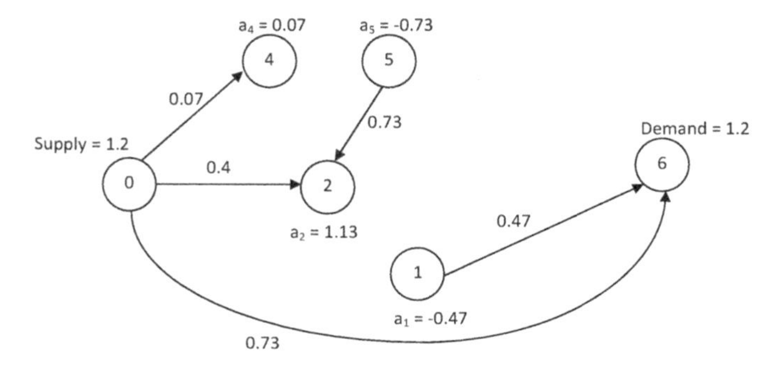

FIGURE 7.9
Network flow problem solution for flow problem shown in Figure 7.8.

We next consider whether the graph $I^{-+} = \{2, 4, 5\}$ can be further subdivided. The network flow subproblem for this subgraph is shown in Figure 7.10, where $\hat{T}^{-+} = \sqrt{13/3}$, and the a_i values have been recomputed and are shown in the figure. The solution that minimizes the flow from the source node to nodes in I^{-+} is shown in Figure 7.11. Note that in this solution, no flow exists from the source node to nodes 2, 4, and 5, indicating that no cut exists with positive value. Because none of the subnetworks may be further divided, this implies that we have an optimal solution for RP. The reorder intervals associated with this solution are shown in Figure 7.12. □

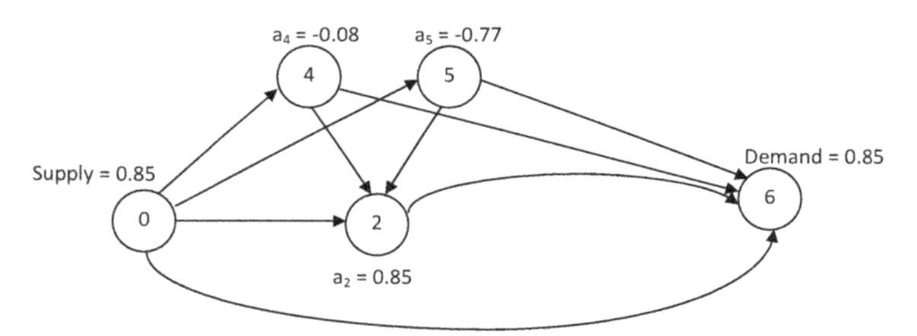

FIGURE 7.10
Network flow problem for I^{-+} resulting from assembly problem example.

Maxwell and Muckstadt [85] showed via induction that the solution obtained via this recursive graph subdivision approach always satisfies $T_i \geq T_j$ for all $(i, j) \in A$, implying that this approach provides a solution that satisfies all of the KKT conditions and, therefore, an optimal solution for RP. Note that solving RP never requires solving more than $|I|$ network flow subproblems, each of which can be solved in $\mathcal{O}(|I|^3)$ time (see [85]).

Having solved problem RP, we now wish to construct a solution for the MSEOQ problem. Maxwell and Muckstadt [85] show that this turns out to

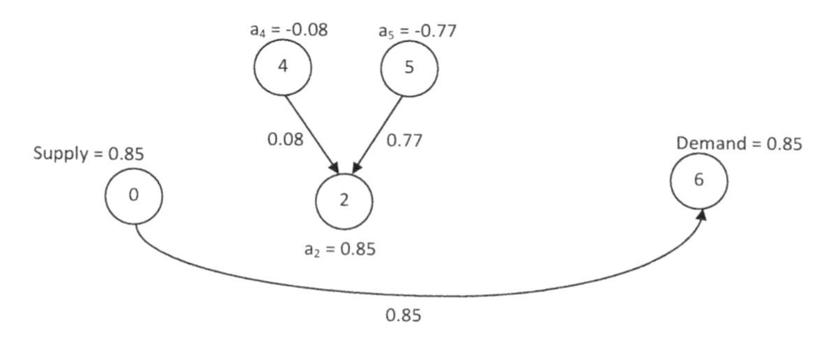

FIGURE 7.11
Network flow problem solution for flow problem shown in Figure 7.10.

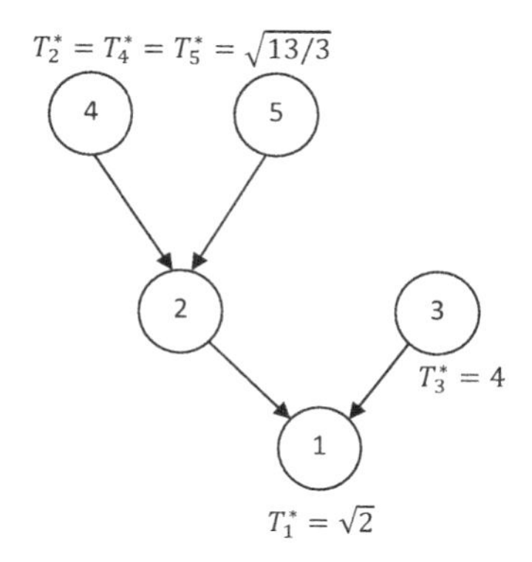

FIGURE 7.12
Optimal reorder intervals for problem RP in assembly problem example.

be very easy to do, although proving that the approach we next describe solves the MSEOQ problem is less than trivial. What Maxwell and Muckstadt show is that independently solving for the best power-of-two multiple for each individual subgraph created in solving RP (assuming that all stages within the same subgraph share the same reorder interval) solves the MSEOQ. Solving for the best power-of-two multiple for a subgraph works as follows. Letting $\tilde{I}$ denote the set of nodes in the subgraph, and because the objective function is convex in the T_i variables, determining the best power-of-two multiple for the set $\tilde{I}$ requires finding the smallest nonnegative integer value of k such that

$$\frac{\sum_{i\in\tilde{I}} f_i}{2^k T_L} + 2^k T_L \sum_{i\in\tilde{I}} g_i \leq \frac{\sum_{i\in\tilde{I}} f_i}{2^{k+1} T_L} + 2^{k+1} T_L \sum_{i\in\tilde{I}} g_i. \tag{7.47}$$

(Note that we implicitly assume here that the reorder interval is at least as great as $2^0 T_L = T_L$.) Because of the way the solution to RP was constructed, Maxwell and Muckstadt [85] show that (i) the optimal solution to the MSEOQ problem can be obtained by solving a relaxed version of the problem in which the optimal solution is determined for each subgraph independently, and (ii) the optimal solution for the restricted version of the problem for a subgraph, in which each node in the subgraph shares a common reorder interval, is also optimal for the subproblem without this requirement. As a result, if the best power-of-two solutions for the individual subgraphs satisfy the nestedness requirement, the combination of these solutions is optimal for the MSEOQ problem. But this occurs as a result of the way the solution to RP was constructed. To see this, note that (7.47) can be rewritten as

$$2^k \geq \frac{1}{\sqrt{2}T_L}\sqrt{\frac{\sum_{i\in\tilde{I}} f_i}{\sum_{i\in\tilde{I}} g_i}}. \tag{7.48}$$

The right-hand side of the above inequality equals a constant multiplied by the reorder interval for the subgraph consisting of the set $\tilde{I}$ in the solution to problem RP. Thus, if $T_i > T_j$ in the solution to RP, then $k_i \geq k_j$ will also hold, where k_i and k_j denote the optimal power-of-two multiples for the subgraphs containing nodes i and j, respectively. For the assembly problem in Example 7.2, the optimal power-of-two solution is given by $k_1 = 0$, $k_3 = 2$, and $k_2 = k_4 = k_5 = 1$, i.e., stage 1 reorders every T_L time units, stage 3 reorders every $4T_L$ time units, and stages 2, 4, and 5 reorder every $2T_L$ time units (where $T_L = 1$ in the example).

Note that the optimal solution value for problem RP serves as a lower bound on the optimal solution value for problem MSEOQ. Suppose that the solution to RP results in n disjoint subgraphs, and let S_l denote the set of nodes in subgraph l, where $\cup_{l=1}^n S_l = I$. Then the optimal reorder interval for subgraph l in problem RP equals $T(S_l) = \sqrt{\sum_{i\in S_l} f_i / \sum_{i\in S_l} g_i}$, and the optimal solution value for RP equals $Z^*_{RP} = 2\sum_{l=1}^n \sqrt{\sum_{i\in S_l} f_i \sum_{i\in S_l} g_i}$. Letting k^*_l denote the optimal power-of-two multiplier in the solution to the

MSEOQ problem, then the optimal cost for this problem equals $Z^*_{MSEOQ} = \sum_{l=1}^{n}\{\sum_{i\in S_l}(f_i/2^{k_l^*}T_L + g_i 2^{k_l^*}T_L)\}$. Observe that by (7.48), k_l^* is the smallest nonnegative integer value of k such that

$$2^k \geq \frac{1}{\sqrt{2}T_L}T(S_l), \tag{7.49}$$

which implies that $2^{k_l^*}T_L \geq T(S_l)/\sqrt{2}$ and $T(S_l)/\sqrt{2} > 2^{k_l^*-1}T_L$, i.e.,

$$\frac{T(S_l)}{\sqrt{2}} \leq 2^{k_l^*}T_L < \sqrt{2}T(S_l), \tag{7.50}$$

assuming $T(S_l) \geq T_L$. Consider the cost of a solution with order replenishment interval lengths equal to the lower or upper limits stated in (7.50). Observe that $\sum_{i\in S_l} f_i/(T(S_l)/\sqrt{2}) + \sum_{i\in S_l} g_i(T(S_l)/\sqrt{2}) = (\sqrt{2}+1/\sqrt{2})\sqrt{\sum_{i\in S_l} f_i \sum_{i\in S_l} g_i}$; similarly, we have $\sum_{i\in S_l} f_i/(\sqrt{2}T(S_l)) + \sum_{i\in S_l} g_i(\sqrt{2}T(S_l)) = (\sqrt{2}+1/\sqrt{2})\sqrt{\sum_{i\in S_l} f_i \sum_{i\in S_l} g_i}$. Condition (7.50) and the convexity of $f_i/T_i + g_iT_i$ therefore imply that $\sum_{i\in S_l}(f_i/2^{k_l^*}T_L + g_i 2^{k_l^*}T_L) \leq (\sqrt{2}+1/\sqrt{2})\sqrt{\sum_{i\in S_l} f_i \sum_{i\in S_l} g_i}$, which implies

$$\begin{aligned} Z^*_{MSEOQ} &= \sum_{l=1}^{n}\left\{\sum_{i\in S_l}\left(f_i/2^{k_l^*}T_L + g_i 2^{k_l^*}T_L\right)\right\} \\ &\leq (\sqrt{2}+1/\sqrt{2})\sum_{l=1}^{n}\sqrt{\sum_{i\in S_l} f_i \sum_{i\in S_l} g_i} \\ &= (\sqrt{2}+1/\sqrt{2})\frac{Z^*_{RP}}{2}. \end{aligned} \tag{7.51}$$

This implies that our power-of-two solution value is no more than $1/\sqrt{2}(1 + 1/2) \approx 1.06$ times the lower bound Z^*_{RP}, i.e., the power-of-two solution is no more than roughly 6% above the optimal solution value among all stationary, nested policies.

Roundy [99] showed this in a more general way and improved upon this 6% worst-case bound by optimizing over the base planning period length T_L. To gain some intuition about this approach, suppose that the optimal solution to problem RP results in only one subgraph, i.e., all stages in the network share the same reorder interval $\hat{T}$. The solution to RP with all $T_i = \hat{T}$ serves as a lower bound on the optimal solution to the MSEOQ problem. Thus, if we could set T_i equal to $\hat{T}$ for all $i \in I$ in the MSEOQ problem, we could do no better than this. This can be achieved by using a base planning period length $T_L = \hat{T}$ (which results in a power-of-two multiple of 0 for each stage). So, while the use of an arbitrary value of T_L guarantees a power-of-two solution with a cost no more than 106% of the optimal solution, if we are free to set the value of T_L, i.e., if T_L is a decision variable, we can improve upon this worst-case performance guarantee.

To show that this is possible, we first define the *effectiveness* of a power-of-two solution as a percentage equal to 100 times the ratio of a lower bound on the optimal solution value to the power-of-two solution value (the lower bound on the solution value equals the solution value of problem RP). The results we have discussed thus far indicate that a power-of-two solution exists with at least 94% effectiveness. Recall that the cost associated with subgraph l in the solution to RP equals $2\sqrt{\sum_{i\in S_l} f_i \sum_{i\in S_l} g_i}$, with an associated reorder interval of $T(S_l)$. Let $M_l/2 = \sum_{i\in S_l} f_i/T(S_l) = \sum_{i\in S_l} g_i T(S_l) = \sqrt{\sum_{i\in S_l} f_i \sum_{i\in S_l} g_i}$, which implies that we can write the optimal solution value for RP as $\sum_{l=1}^{n} M_l$. For any value of the reorder interval for subgraph l, T_l, define $q_l = T(S_l)/T_l$ and let $e(q_l) = 2/(q_l + 1/q_l)$. Using these definitions, the cost associated with subgraph l when using a reorder interval of T_l equals

$$\frac{\sum_{i\in S_l} f_i}{T_l} + \sum_{i\in S_l} g_i T_l = \frac{q_l \sum_{i\in S_l} f_i}{T(S_l)} + \sum_{i\in S_l} g_i T(S_l)/q_l = \frac{M_l}{e(q_l)}, \qquad (7.52)$$

where we have used the fact that $e(q_l) = e(1/q_l)$. The above equation and the definition of M_l as the minimum subgraph l cost imply that the *effectiveness* of the solution for an individual subgraph l equals $M_l/(M_l/e(q_l)) = e(q_l)$.

By definition, the effectiveness of a solution in which the reorder interval for subgraph l equals T_l is equal to

$$\frac{Z_{RP}^*}{\sum_{l=1}^{n} \frac{M_l}{e(q_l)}} = \frac{\sum_{l=1}^{n} M_l}{\sum_{l=1}^{n} \frac{M_l}{e(q_l)}}. \qquad (7.53)$$

Because $\sum_{l=1}^{n}(M_l/e(q_l)) \leq \left(\sum_{l=1}^{n} M_l\right)/[\min_{l=1,\ldots,n}\{e(q_l)\}]$, we have

$$\frac{Z_{RP}^*}{\sum_{l=1}^{n} \frac{M_l}{e(q_l)}} \geq \min_{l=1,\ldots,n}\{e(q_l)\}\frac{\sum_{l=1}^{n} M_l}{\sum_{l=1}^{n} M_l} = \min_{l=1,\ldots,n}\{e(q_l)\}. \qquad (7.54)$$

Equation (7.54) provides a lower bound on the effectiveness of a policy, equal to the minimum value of the effectiveness among all subgraphs.

The function $e(q_l)$ is unimodal with a maximum at $q_l = 1$, and has the property that $e(q_l) = e(1/q_l)$. Noting that $e(\sqrt{2}) = 0.94$, we will have a policy with at least 94% effectiveness if $1/\sqrt{2} \leq q_l \leq \sqrt{2}$ for all $l = 1, \ldots, n$, which is what the method we described for solving the MSEOQ problem ensures, as this condition is equivalent to (7.50). Thus, when the ratio of $T(S_l)$ to the power-of-two reorder interval for subgraph l is in the interval $[1/\sqrt{2}, \sqrt{2}]$, we ensure at least a 94% effective solution. The power-of-two reorder interval for subgraph l is stated in terms of the base planning period length, T_L, and we would like to optimize over the value of T_L; because we are considering power-of-two multiples of the base planning period length, we do not lose anything by restricting our attention to values of T_L on the interval $[1/\sqrt{2}, \sqrt{2}]$ (any power-of-two multiple of a positive base planning period length outside this interval can be expressed as a power-of-two multiple of a base planning period length within this interval; Exercise 7.6 at the end of the chapter asks the reader to prove this claim).

Next, consider what happens to the optimal power-of-two multiple for subgraph l as T_L increases from $1/\sqrt{2}$ to $\sqrt{2}$. Let $T_2^l(T_L)$ denote the optimal power-of-two reorder interval for subgraph l as a function of the base planning period length T_L. Then by (7.50) we have $T_2^l(T_L) = 2^{k_l} T_L$ when

$$2^{-k_l-1/2} \leq \frac{T_L}{T(S_l)} < 2^{-k_l+1/2}. \tag{7.55}$$

Suppose some k_l' satisfies (7.55) at $T_L = 1/\sqrt{2}$. If we increase T_L from $1/\sqrt{2}$ but maintain $k_l = k_l'$, then the right-hand side (strict) inequality will become violated at $T_L = 2^{-k_l'+1/2}T(S_l) > 1/\sqrt{2}$; at this value of T_L, the value of k_l satisfying (7.55) will equal $k_l' - 1$. If we were to continue increasing T_L while maintaining k_l fixed at $k_l' - 1$, then the next point at which (7.55) would be violated would occur at $T_L = 2^{-k_l'+1/2+1}T(S_l) > 2/\sqrt{2} = \sqrt{2}$. Thus, as we increase T_L from $1/\sqrt{2}$ to $\sqrt{2}$, the optimal multiplier value k_l changes only once at the point $T_L = 2^{-k_l'+1/2}T(S_l)$, at which point it decreases by one. Because this change occurs once for each of the n subgraphs, we need only consider $n + 1$ values of the vector $\{k_1, k_2, \ldots, k_n\}$ of multiplier values. For subgraph $l = 1, \ldots, n$, let $\delta_l = 2^{-k_l'+1/2}T(S_l)$ denote the value of T_L on the interval $[1/\sqrt{2}, \sqrt{2}]$ such that the optimal multiplier for the subgraph changes from k_l' to $k_l' - 1$. Assume without loss of generality that the subgraphs are re-indexed in increasing order of δ_l values. Defining $\delta_0 = 1/\sqrt{2}$ and $\delta_{n+1} = \sqrt{2}$, then we have $n+1$ contiguous intervals on $[1/\sqrt{2}, \sqrt{2}]$, (δ_{l-1}, δ_l), $l = 1, \ldots, n+1$, such that the optimal power-of-two multiple vector is invariant within the interval.

For a given value of T_L on the interval (δ_{l-1}, δ_l), subgraphs l through n have optimal multiplier values equal to k_l', while subgraphs 1 through $l - 1$ have optimal multiplier values equal to $k_l' - 1$. The cost of the solution in T_L associated with the optimal power-of-two multiples for this interval equals

$$\sum_{r=1}^{l-1} \sum_{i \in S_r} \left(\frac{f_i}{2^{k_r'-1} T_L} + g_i 2^{k_r'-1} T_L \right) + \sum_{r=l}^{n} \sum_{i \in S_r} \left(\frac{f_i}{2^{k_r'} T_L} + g_i 2^{k_r'} T_L \right). \tag{7.56}$$

This can be written more compactly by letting $f_l' = \sum_{r=1}^{l-1} \sum_{i \in S_r} f_i / 2^{k_r'-1} + \sum_{r=l}^{n} \sum_{i \in S_r} f_i / 2^{k_r'}$ and $g_l' = \sum_{r=1}^{l-1} \sum_{i \in S_r} g_i 2^{k_r'-1} + \sum_{r=l}^{n} \sum_{i \in S_r} g_i 2^{k_r'}$, in which case (7.56) can be written as

$$\frac{f_l'}{T_L} + g_l' T_L, \tag{7.57}$$

and the optimal value of the base planning period length associated with the corresponding set of power-of-two multiples is given by $T_L = \sqrt{f_l'/g_l'}$, with average annual cost $2\sqrt{f_l'g_l'}$. As a result, the best choice of l is the one that produces the minimum value of $f_l'g_l'$.

Roundy [99] uses an extremely clever method to demonstrate the effectiveness of a power-of-two policy that uses an optimal base period length. Using earlier notation, we can write the cost of the power-of-two policy as $\sum_{l=1}^{n} M_l/e(q_l)$, where $q_l = T(S_l)/T_l$ and $e(q_l) = 2/(q_l + 1/q_l)$. We would like to characterize the minimum value of $\sum_{l=1}^{n} M_l/e(q_l)$ when using the optimal base planning period length T_L on the interval $[1/\sqrt{2}, \sqrt{2}]$. Roundy [99] considers the expected value of the cost $\sum_{l=1}^{n} M_l/e(q_l)$ assuming a distribution of T_L on the interval $[1/\sqrt{2}, \sqrt{2}]$ according to the density function $1/(T_L \ln 2)$. This expected value serves as an upper bound on the minimum cost value on the interval. To compute this expected value, we need to compute the expected value of each $1/e(q_l)$ term on the interval, as $E[M_l/e(q_l)] = M_l E[1/e(q_l)]$. Roundy [99] shows that when using the chosen density function, then $E[1/e(q_l)] = 1/(\sqrt{2}\ln 2)$,[7] which implies that the expected cost of the solution equals $\sum_{l=1}^{n} M_l/(\sqrt{2}\ln 2) = Z_{RP}^*/(\sqrt{2}\ln 2)$. The effectiveness of this policy therefore equals $100\% \times \sqrt{2}\ln 2 > 98\%$.

7.3.2 The joint replenishment problem

Although the joint replenishment problem (JRP) does not consider multiple locations or production stages, it can be modeled as an equivalent multistage assembly problem, as we next discuss. The JRP considers a set I of different products, such that item i has a constant annual demand rate equal to λ_i and a holding cost rate of h_i. We incur a fixed cost of f_i each time we place a replenishment order for item i; in addition, whenever we order any item, a "major" fixed order cost of f_0 is also incurred. This additional order cost corresponds to a fixed cost that products may share. Suppose we model each item as a production stage, with a corresponding node in a multistage production network. We can model the shared fixed order cost as a production stage, denoted as stage 0, which has an associated order cost but no inventory cost; this stage must order each time any other stage orders. Thus, if we create

[7] Observe that if $T_L \le \delta_l$, then $q_l = \delta_l/\sqrt{2}T_L$; if $T_L \ge \delta_l$, then $q_l = T_L/\sqrt{2}\delta_l$. Similarly, if $T_L \le \delta_l$ then $1/e(q_l) = \delta_l/2\sqrt{2}T_L + T_l/\sqrt{2}\delta_l$ and if $T_L \ge \delta_l$ then $1/e(q_l) = T_l/2\sqrt{2}\delta_l + \delta_l/\sqrt{2}T_L$. Thus,

$$
\begin{aligned}
E\left[\frac{1}{e(q_l)}\right] &= \int_{\frac{1}{\sqrt{2}}}^{\sqrt{2}} \frac{1}{e(q_l)} \frac{dT_L}{T_L \ln 2} \\
&= \int_{\frac{1}{\sqrt{2}}}^{\delta_l} \left(\frac{\delta_l}{2\sqrt{2}T_L^2} + \frac{\sqrt{2}}{2\delta_l}\right)\frac{dT_L}{\ln 2} + \int_{\delta_l}^{\sqrt{2}} \left(\frac{1}{2\sqrt{2}\delta_l} + \frac{\delta_l}{\sqrt{2}T_L^2}\right)\frac{dT_L}{\ln 2} \\
&= \frac{1}{\sqrt{2}\ln 2}.
\end{aligned}
$$

an arc from each product stage node to the shared fixed order cost stage 0, and require the use of a nested policy, this will force stage 0 to order each time any product stage orders. The resulting network is a two-level assembly network in which stages 1 through n exist at the upper level, stage 0 exists at the lower level, and an arc exists from stage i to 0 for $i = 1, \ldots, n$. Let $g_i = h_i \lambda_i / 2$, $i = 0, \ldots, n$, where $h_0 = 0$. We can, therefore, apply the methods discussed in the previous subsection to provide a power-of-two policy with 98% effectiveness.

Consider what happens, however, when we apply the algorithm for decomposing the network into subgraphs in the previous subsection to the JRP in order to solve the relaxation. At the first iteration in solving the relaxed problem, we obtain a value of $\hat{T} = \sqrt{\sum_{i=0}^{n} f_i / \sum_{i=0}^{n} g_i}$. Because $g_0 = 0$, node 0 will always have a value of $a_0 = f_0/\hat{T}^2$, which is positive, implying that node 0 will be a net-demand node in all of the network flow subproblems. Given a value of $\hat{T}$ for a subgraph, a stage i such that $a_i = f_i/\hat{T}^2 - g_i > 0$, i.e., $\sqrt{f_i/g_i} > \hat{T}$, has a positive partial derivative of cost with respect to T_i at $\hat{T}$. Such nodes have net demand $a_i > 0$ in the network flow subproblem. A stage i such that $a_i \leq 0$, i.e., $\sqrt{f_i/g_i} \leq \hat{T}$, has a nonpositive partial derivative of cost with respect to T_i at $\hat{T}$. Such nodes have net supply $-a_i \geq 0$ in the network flow subproblem. Suppose, without loss of generality, that we index items in nondecreasing order of f_i/g_i. Then, for some index $i = \hat{\imath}$, the nodes from $i = 1, \ldots, \hat{\imath}$ have nonnegative net supply, while those nodes from $\hat{\imath} + 1, \ldots, n$ have positive net demand.

Note that in the network flow subproblem, those nodes with index $\hat{\imath} + 1$ or greater correspond to net demand nodes, but they cannot receive flow from anywhere but the dummy source node. Moreover, the sum of the supplies from nodes 1 through $\hat{\imath}$ must be greater than or equal to the net demand at node 0, which implies that the subgraph containing nodes $0, 1, \ldots, \hat{\imath}$ will have net excess supply and will form the set I^-, while nodes $\hat{\imath}, \ldots, n$ will have net excess demand and will form the set I^+. Once the subgraph consisting of nodes $\hat{\imath}, \ldots, n$ is disconnected from node 0, repeated application of the subgraph division network flow problem will result in each node in this group forming a subgraph of one node, i.e., each node $i \in I^+$ will ultimately have a reorder interval equal to $\sqrt{f_i/g_i}$ in the optimal relaxation solution (Exercise 7.8 asks the reader to prove this result). Restarting the subgraph division with $I = I^-$ will produce the same type of structural result, i.e., any node $i \in I^{-+}$ will have an optimal reorder interval of $\sqrt{f_i/g_i}$ in an optimal relaxation solution. The subgraph division algorithm will terminate when, for some subgraph I^-, all nodes in $I^- \backslash \{0\}$ have net supply, in which case the optimal reorder interval for all nodes in I^- (which includes node 0) equals $\sqrt{\sum_{i \in I^-} f_i / \sum_{i \in I^-} g_i}$. As a result, the structure of the optimal relaxation solution is given by

$$T_i^* = \begin{cases} \sqrt{\frac{\sum_{i=0}^{\hat{\imath}} f_i}{\sum_{i=1}^{\hat{\imath}} g_i}}, & i = 0, \ldots, \hat{\imath}, \\ \sqrt{f_i/g_i}, & i = \hat{\imath}+1, \ldots, n, \end{cases} \tag{7.58}$$

for some value of $\hat{\imath} = 1, \ldots, n$. Thus, there is no need to solve the subgraph division network flow problem at all. Instead we can simply sort items in nondecreasing order of f_i/g_i and compute the cost of each solution of the form of (7.58) in $\mathcal{O}(n \log n)$ time. The power-of-two solution obtained from this relaxation solution will have 94% effectiveness for an arbitrary base planning period length, and 98% effectiveness when the base planning period length is optimized, as described in the previous subsection.

7.3.3 The one-warehouse multi-retailer problem

As noted earlier, the class of nested policies may perform arbitrarily poorly for distribution systems (see [98]), within which class the one-warehouse multi-retailer problem falls. This section analyzes this special class of distribution systems in a similar manner as in the previous two subsections, but without the requirement of a nested policy structure. That is, we confine our analysis to the class of stationary power-of-two policies in distribution systems in which a single warehouse supplies n different retailers with a product, and where the demand seen by retailer i occurs at a rate of λ_i units per year for $i = 1, \ldots, n$. The associated network consists of a node 0, corresponding to the warehouse, n retailer nodes, and an arc from node 0 to node i, for $i = 1, \ldots, n$, for a total of n arcs (thus, this network is an inverted version of the network structure for the JRP in the previous subsection). As in Subsection 7.3.1, we let h_i denote the echelon holding cost rate at node i, for $i = 0, 1, \ldots, n$, with $g_i = h_i\lambda_i/2$ for $i = 1, \ldots, n$. At the warehouse level, because the demand occurs at a constant, deterministic rate, it is convenient to keep track of the echelon holding cost associated with each retailer separately. Let $g_{0i} = h_0\lambda_i/2$ for $i = 1, \ldots, n$. To compute the average echelon inventory level, we first consider the case in which retailer i's reorder interval is greater than or equal to the warehouse reorder interval, i.e., $T_i \geq T_0$. When $T_i \geq T_0$, the warehouse orders more frequently than retailer i and because we are using a power-of-two policy, this implies that the warehouse places an order every time retailer i does, and immediately ships all inventory ordered to the retailer, holding no inventory for this retailer. This implies that the echelon inventory level at the warehouse for retailer i is the same as the local inventory level at retailer i. Thus, the average echelon holding cost at the warehouse associated with retailer i equals $g_{0i}T_i$. When $T_0 \geq T_i$, we are back in the nested policy case, which implies that the average echelon holding cost at the warehouse associated with retailer i equals $g_{0i}T_0$. This implies that the warehouse average echelon holding cost associated with retailer i equals $g_{i0}\max\{T_i, T_0\}$, and the average annual system holding cost associated with retailer i inventory can be expressed as $g_iT_i + g_{i0}\max\{T_i, T_0\}$.

The average annual cost associated with a stationary power-of-two policy thus equals

$$\sum_{i=0}^{n} \frac{f_i}{T_i} + \sum_{i=1}^{n} \{g_i T_i + g_{i0} \max\{T_i, T_0\}\}. \quad (7.59)$$

We wish to minimize this function over all vectors $T = \{T_0, T_1, \ldots, T_n\}$ such that each element of this vector takes a value equal to a power-of-two multiple of a base planning period length T_L. That is, we wish to solve

[OWMR]

$$\begin{aligned} \text{Minimize} \quad & \textstyle\sum_{i=0}^{n} \frac{f_i}{T_i} + \sum_{i=1}^{n} \{g_i T_i + g_{i0} \max\{T_i, T_0\}\} && (7.60) \\ \text{Subject to:} \quad & T_i = 2^{k_i} T_L, \qquad i = 0, \ldots, n, && (7.61) \\ & k_i \in \mathbb{Z}_+, \qquad i \in I. && (7.62) \end{aligned}$$

As in Subsection 7.3.1, we begin by considering the relaxation of this problem in which the power-of-two restriction is relaxed. That is, we replace the constraints with the simple nonnegativity constraints $T_i \geq 0$, $i = 0, \ldots, n$. For any given $T_0 = \hat{T}_0$, we can easily determine the optimal value of T_i for $i > 0$, as well as the associated minimum cost when $T_0 = \hat{T}_0$. Doing so requires solving problem SP_i:

$$[SP_i] \qquad \begin{aligned} \text{Minimize} \quad & \tfrac{f_i}{T_i} + g_i T_i + g_{i0} \max\{T_i, \hat{T}_0\} && (7.63) \\ \text{Subject to:} \quad & T_i \geq 0. && (7.64) \end{aligned}$$

Suppose we minimize SP_i over all $T_i \leq \hat{T}_0$. The optimal value of T_i over this region is given by $\min\{\hat{T}_0, \sqrt{f_i/g_i}\}$. Next, suppose we minimize SP_i over all $T_i \geq \hat{T}_0$. In this case, the optimal value of T_i over this region is given by $\max\{\sqrt{f_i/(g_i + g_{i0})}, \hat{T}_0\}$. Observe that $\overline{T}_i \equiv \sqrt{f_i/g_i} \geq \sqrt{f_i/(g_i + g_{i0})} \equiv \underline{T}_i$, and one of the conditions $\hat{T}_0 < \underline{T}_i$, $\overline{T}_i \geq \hat{T}_0 \geq \underline{T}_i$, or $\hat{T}_0 > \overline{T}_i$ must hold. When $\hat{T}_0 < \underline{T}_i$, then $\underline{T}_i$ must be an optimal solution. When $\overline{T}_i \geq \hat{T}_0 \geq \underline{T}_i$, then $\hat{T}_0$ must be an optimal solution, and when $\hat{T}_0 > \overline{T}_i$, then $\overline{T}_i$ must be an optimal solution. As a result, when $\hat{T}_0 < \underline{T}_i$, the optimal SP_i cost equals $2\sqrt{f_i(g_i + g_{i0})}$. When $\overline{T}_i \geq \hat{T}_0 \geq \underline{T}_i$, the optimal SP_i cost equals $f_i/\hat{T}_0 + (g_i + g_{i0})\hat{T}_0$, and when $\hat{T}_0 > \overline{T}_i$, the optimal SP_i cost equals $2\sqrt{f_i g_i} + g_{i0}\hat{T}_0$.

Let $b_i(T_0)$ equal the optimal SP_i cost as a function of T_0; determining the optimal value of T_0 requires minimizing $B(T_0) = f_0/T_0 + \sum_{i=1}^{n} b_i(T_0)$, where

$$b_i(T_0) = \begin{cases} 2\sqrt{f_i(g_i + g_{0i})}, & T_0 < \underline{T}_i, \\ \frac{f_i}{T_0} + (g_i + g_{i0})T_0, & \underline{T}_i \leq T_0 \leq \overline{T}_i, \\ 2\sqrt{f_i g_i} + g_{i0} T_0, & \overline{T}_i < T_0. \end{cases} \quad (7.65)$$

Observe that the functional form of $b_i(T_0)$ can only change at the points

$\underline{T}_i$ and $\overline{T}_i$, for $i = 1, \ldots, n$, and one can easily verify that $b_i(T_0)$ is a continuously differentiable and convex function of T_0. Thus, $B(T_0)$ is a convex and continuously differentiable function; in fact, as the sum of this set of continuously differentiable convex functions plus the strictly convex function f_0/T_0, $B(T_0)$ is a strictly convex and continuously differentiable function. Moreover, $B(T_0) \to \infty$ as $T_0 \to \infty$ and as T_0 tends to zero from above, implying that $B(T_0)$ has a unique positive stationary point optimal solution. $B(T_0)$ changes functional form at each $\underline{T}_i$ and $\overline{T}_i$ value, $i = 1, \ldots, n$, and the set of all $\underline{T}_i$ and $\overline{T}_i$ values, when sequenced from smallest to largest, forms a set of consecutive intervals on the real line. Within any such interval, $B(T_0)$ takes the form $f(T_0)/T_0 + c(T_0) + g(T_0)T_0$, where the values of $f(T_0)$, $c(T_0)$ and $g(T_0)$ are constants within the interval. Letting r index these $2n+1$ intervals, then the cost equation associated with interval r may be written as $f_r(T_0)/T_0 + c_r(T_0) + g_r(T_0)T_0$; the stationary point solution associated with this function is given by $T_0^r = \sqrt{f_r(T_0)/g_r(T_0)}$. For each interval, we can compute T_0^r; because $B(T_0)$ is strictly convex and continuously differentiable, this implies that if the value of T_0^r falls within the rth interval, then this value of T_0^r is the unique global minimizer of $B(T_0)$. This implies that minimizing $B(T_0)$ requires computing the value of T_0^r for each of the $2n+1$ intervals, and comparing this value to the associated interval bounds. A value of T_0^r that is feasible for its interval bounds gives an optimal solution.

At the optimal value of $T_0 = T_0^*$, we can determine the optimal value of T_i based on (7.65). If $T_0 < \underline{T}_i$, then $T_i^* = \underline{T}_i$. Let G denote the set of retailers satisfying this condition, i.e., whose optimal reorder intervals are Greater than T_0^*. Similarly, let E denote the set of retailers such that $\underline{T}_i \le T_0 \le \overline{T}_i$, i.e., whose optimal reorder intervals are Equal to T_0^*, and let L denote the set of retailers with $\overline{T}_i < T_0$, i.e., whose optimal reorder intervals are Less than T_0^*. The average annual cost under these set definitions can then be written as

$$\frac{\tilde{f}}{T_0} + \tilde{g}T_0 + \sum_{i \in L \cup G} \left\{ \frac{f_i}{T_i} + \tilde{g}_i T_i \right\}, \tag{7.66}$$

where $\tilde{f} = f_0 + \sum_{i \in E} f_i$, $\tilde{g} = \sum_{i \in E}(g_i + g_{i0}) + \sum_{i \in L} g_{i0}$, and $\tilde{g}_i = g_i + g_{i0}$ if $i \in G$, while $\tilde{g}_i = g_i$ if $i \in L$. The optimal relaxation solution can then be written as $T_0^* = \sqrt{\tilde{f}/\tilde{g}} = T_i^*$ for $i \in E$, and $T_i^* = \sqrt{f_i/\tilde{g}_i}$ for $i \in L \cup G$. Roundy [98] referred to any solution vector $\vec{T}$ such that all members of the set E have the same reorder interval T_0, each member of G has a reorder interval greater than or equal to T_0, and each member of L has a reorder interval less than or equal to T_0 as an *order-preserving* solution. Observe that the cost of any order-preserving policy may be computed using (7.66).

The value of (7.66) in the optimal relaxation solution may be written as $Z_R^* = M_0 + \sum_{i \in L \cup G} M_i$, where $M_0 = 2\sqrt{\tilde{f}\tilde{g}}$ is the value of the sum of the first two terms in (7.66) at optimality and $M_i = 2\sqrt{f_i \tilde{g}_i}$ is the ith element of the summation at optimality. Roundy [98] showed that Z_R^* provides a lower bound on the cost of *any feasible* policy. Showing the effectiveness of an optimal power-of-two policy uses essentially the same approach discussed in Subsection 7.3.1. Observe that any function of the form $f/T_i + gT_i$, which is minimized at $T_i^* = \sqrt{f/g}$, can be written as $(f/T_i^* + gT_i^*)(1/e(q_i))$, where $e(q_i) = 2/(q_i + 1/q_i)$ and $q_i = T_i^*/T_i$. Because of this, and by the definition of M_i, $i = 0, \ldots, n$, we can write (7.66) for any order-preserving solution as equal to

$$\frac{M_0}{e(q_0)} + \sum_{i \in L \cup G} \frac{M_i}{e(q_i)}. \tag{7.67}$$

As in Equation (7.54), we can write the effectiveness of an order-preserving solution as 100% times

$$\frac{Z_R^*}{\frac{M_0}{e(q_0)} + \sum_{i \in L \cup G} \frac{M_i}{e(q_i)}} \geq \min_{i=0,\ldots,n}\{e(q_i)\} \frac{Z_R^*}{M_0 + \sum_{i \in L \cup G} M_i} = \min_{i=0,\ldots,n}\{e(q_i)\}. \tag{7.68}$$

Thus, the minimum effectiveness at a stage (where stage 0 includes all $i \in E$) provides a lower bound on the effectiveness of an order-preserving solution. Subsection 7.3.1 showed the effectiveness of a power-of-two policy (94% for an arbitrary base planning period length and 98% if we optimize over the base planning period length), and these results apply to the OWMR problem, provided that the power-of-two solution is order preserving. The power-of-two solution must, however, be order preserving, as the smallest nonnegative value of k such that $2^k \geq T_i^*/\sqrt{2}T_L$ cannot be larger than the smallest nonnegative value of k such that $2^k \geq T_0^*/\sqrt{2}T_L$ if $T_i^* < T_0^*$, which holds for $i \in L$. Similarly, the value smallest nonnegative value of k such that $2^k \geq T_i^*/\sqrt{2}T_L$ cannot be smaller than the smallest nonnegative value of k such that $2^k \geq T_0^*/\sqrt{2}T_L$ if $T_i^* > T_0^*$, which holds for $i \in G$.

Example 7.3 Consider a one-warehouse multi-retailer system with $n = 5$ retailers and $T_L = 1$. The costs and demand rates associated with each stage are provided in the table below (note that stage 0 corresponds to the warehouse).

Stage, i	0	1	2	3	4	5
λ_i		100	200	1000	500	2000
g_i		5	10	50	25	100
g_{0i}		50	100	500	250	1000
f_i	500	500	500	100	250	250

The table below provides the values of $\underline{T}_i$ and $\overline{T}_i$, which defines a set of 11 intervals for the value of T_0, such that the cost function associated with an interval is different for each interval.

Stage i	1	2	3	4	5
$\underline{T}_i$	3.02	2.13	0.43	0.95	0.48
$\overline{T}_i$	10.00	7.07	1.41	3.16	1.58

The optimal relaxation solution occurs on the interval (0.48, 0.95) at $T_0^* = 0.72$, with a cost of 3693.65. The set E contains stages 3 and 5, the set G contains stages 1, 2, and 4, and the set L is empty. This solution sets $T_0^* = T_3^* = T_5^* = 0.72$, $T_1^* = 3.02$, $T_2^* = 2.13$, and $T_4^* = 0.95$. The power-of-two multiples for each stage, obtained by finding the smallest k such that $2^k \geq T_i^*/\sqrt{2}T_L$, equal 1, 2, 2, 1, 1, and 1, respectively, for stages 0 through 5. The optimal power-of-two rounded solution has solution value 3855, giving an effectiveness of 0.958. Observe that if we set our base planning period length equal to $T_0^* = 0.72$, the resulting power-of-two solution sets $T_0 = T_3 = T_4 = T_5 = 0.72$ and $T_1 = T_2 = 2.87$, with a total cost of 3736.26, and an effectiveness of 0.989. □

We end this section by noting that a stream of work exists that analyzes this class of multistage production systems in the presence of some form of production capacity limits. Jackson, Maxwell, and Muckstadt [67] considered multistage problems involving workcenters, where each workcenter has a limit on the total amount of time that may be spent performing setups in a year, and where a setup at a production stage within a workcenter takes a fixed amount of time. A workcenter consists of a subgraph of the overall multistage production system graph, and they assume that workcenters do not share production stages. They seek a power-of-two solution and begin by solving a relaxed problem in which the power-of-two requirement is omitted and the capacity constraints are retained. This problem is solved via Lagrangian relaxation for a single workcenter, and the authors show that the cost of a power-of-two rounded solution does not exceed 141% of the cost of the optimal relaxation solution, and can exceed the capacity limits by no more than 41% as well. Roundy [100] subsequently provided a power-of-two roundoff algorithm for this problem that guarantees capacity feasibility and provides a cost increase of no more than 44% above the cost of an optimal solution. Federgruen and Zheng [44] subsequently considered the general acyclic graph problem defined by Maxwell and Muckstadt [85] and Roundy [99], but with the addition of bounds on the reorder interval values at all stages (which may represent various types of capacity limits). They provide an exact algorithm for optimal power-of-two reorder intervals for this capacitated problem version, with complexity that is comparable to the best algorithmic approaches for the uncapacitated case.

7.4 Review

This chapter considered a variety of multistage production and distribution planning problems. We began by considering dynamic, discrete time models in which demands and costs may vary from period to period. Section 7.2.1 provided a polynomial time dynamic programming solution approach for the serial system structure. Section 7.2.2 then considered more general network structures, and the use of Lagrangian relaxation as a method for obtaining tight bounds on the optimal solution value. We next discussed constant-factor approximations for the dynamic one-warehouse multi-retailer (OWMR) problem, results which also extend to the dynamic joint replenishment problem (JRP). Section 7.3 presented multistage problems based on the economic order quantity (EOQ) model assumptions, and considered the important class of power-of-two policies and their effectiveness in obtaining near-optimal solution values under various network structures. Sections 7.3.2 and 7.3.3 covered the constant demand rate version of the specially structured JRP and OWMR problems, respectively.

Exercises

Ex. 7.1 — Provide a formal proof that a nested production schedule is optimal for the SULS problem in Section 7.2.1 under nonincreasing production costs over time at each stage and nondecreasing holding costs in the stage index within each time period (without referring to the paper by Love [81]!).

Ex. 7.2 — Consider the restricted version of the DOWMR formulation (7.27)–(7.31) in which the y_{0t} variables are fixed to binary values (with $y_{01} = 1$). Using what we learned in Chapter 5, explain why we can be sure that an optimal solution exists to the LP relaxation of this restricted problem that satisfies the binary requirements on the y_{jt} variables for all $j \in J$ and $t = 1, \ldots, T$.

Ex. 7.3 — For a two-stage serial system, verify that the sum of the product of the echelon stock levels and echelon holding cost rates equals the sum of the product of installation holding cost levels and installation holding cost rates.

Ex. 7.4 — Write out all of the KKT conditions for Problem RP when the corresponding network is a serial system with n stages.

Ex. 7.5 — Suppose that for problem RP, we begin with a set of nodes I such that in the optimal solution, not all items share the same reorder interval, i.e., the solution to the network flow problem for determining the cut with the maximum value requires flow from node 0 to at least one original node in the network. Show that we cannot have arcs directed from the subnetwork with excess supply to the subnetwork with excess demand in the optimal solution to the network flow subproblem for finding the cut with the maximal value.

Ex. 7.6 — Show that any power-of-two multiple of a positive base planning period length outside the interval $[1/\sqrt{2}, \sqrt{2}]$ can be expressed as a power-of-two multiple of a base planning period length within the interval $[1/\sqrt{2}, \sqrt{2}]$.

Ex. 7.7 — Verify the expected value computation for $1/e(q_l)$ in Footnote 7 on Page 128.

Ex. 7.8 — Show that in solving the JRP relaxation problem, after the first application of the subgraph division via the network flow subproblem, each node $i \in I^+$ will have an optimal reorder interval equal to $\sqrt{f_i/g_i}$.

Ex. 7.9 — Show that for problem SP_i in Section 7.3.3, when $T_0 < \underline{T}_i$, then $\underline{T}_i$ must be an optimal solution; when $\overline{T}_i \geq T_0 \geq \underline{T}_i$, then T_0 must be an optimal solution; and when $T_0 > \overline{T}_i$, then $\overline{T}_i$ must be an optimal solution.

8

Discrete Facility Location Problems

8.1 Introduction

The field of location theory is both broad and deep, and the research literature in this area is voluminous. Our intent in this chapter is to cover a specific segment of this literature that is particularly relevant to strategic operations planning. The key feature of the location model we discuss in this chapter lies in the fact that, in its most general form, this model generalizes each of the models we have discussed thus far, with the exception of the multistage EOQ models of Chapter 7 and the set covering, packing, and partitioning problems in Chapter 3 (the uncapacitated and single-sourcing location problems we discuss may actually be formulated as special cases of the set packing or partitioning formulations). Thus, our goals in this chapter are twofold. The first of these is to highlight the close relationships between discrete location problems and the previous models covered in this book. The second of these goals is to understand effective solution methods for discrete facility location problems.

The study of discrete facility location models is motivated by the need to meet a discrete set of demands over a long-term time horizon (e.g., multiple years). These demands must be met by supply points, which may correspond to factories or warehouses, for example. The capital investment in such supply points is typically large, with the associated expense incurred over multiple years. The location models we will discuss do not, however, have an explicit time dimension. Instead, we typically have in mind an implicit time dimension associated with the problem; for example, we may be interested in the annual cost associated with a potential solution, assuming the parameters used in the model are representative of a typical year. Clearly this modeling approach may omit important practical problem features, such as the change in costs and demands from year-to-year as well as uncertainty in problem data (the same may be said for all of the models discussed in this book). As a result, such deterministic models typically serve as an approximation of the actual problem and will rely on effective parameter forecasting, estimation, and sensitivity analysis before implementation.

It is convenient to think of this set of demands as associated with a discrete set of customers or markets. Thus, if I denotes a discrete and finite set of customers, we let d_i denote the demand in customer i for each $i \in I$ over a time period for a single product (e.g., annual demand). In addition to this set of demands, we have a discrete set of *candidate* supply points J. For any particular candidate location $j \in J$, we may choose to activate, or operate, this location as a supply point (this decision to activate a supply point may correspond to the construction of a factory or warehouse, or to the ongoing operation of an already existing supply point). Associated with the operation of a supply point is a fixed (e.g., annualized) cost, which corresponds to the cost incurred independent of the facility's output. Let f_j denote this fixed cost for $j \in J$. To keep track of whether or not this cost is incurred for each supply point j, we create a binary variable y_j, which equals one if facility $j \in J$ is active, and equals zero otherwise. If supply point $j \in J$ sends x_{ij} units to customer $i \in I$, then a variable cost equal to $c_{ij}x_{ij}$ is incurred, where c_{ij} denotes the cost for sending a unit from supply point j to customer i. This parameter c_{ij} must, therefore, account for all variable production, holding, and transportation costs incurred in supplying customer i from facility j (as a result, the model implicitly assumes that any fixed costs associated with production and transportation are allocated across all units annually).

We will begin the discussion of discrete facility location models by formulating a general profit-maximizing version of the model, wherein the supplier may choose not only which candidate facilities to operate and which customers are served by which supply points, but also whether or not to serve a customer. After this, we will discuss the more standard cost-minimizing models in which the supplier must meet all customer demands. The following section demonstrates how the class of discrete facility location problems we consider is related to the previous problems covered in this book. Later in this chapter, we consider effective solution methods for classical discrete facility location problems in both the uncapacitated and capacitated cases.

8.2 Relation to Previous Models in this Book

In the profit-maximizing version of the discrete facility location problem (FLP), we allow the supplier to determine whether or not to satisfy some or all of each customer's demand. Let z_i denote the number of units of customer i demand that the supplier chooses to satisfy (where we require $z_i \in [0, d_i]$ for each $i \in I$). Thus, under this interpretation, the demand quantity, d_i, for customer $i \in I$, corresponds to the maximum amount of customer i demand that the supplier may choose to satisfy. Associated with each customer $i \in I$ is a per-unit revenue equal to r_i, which implies that the total revenue obtained from customer i will equal $r_i z_i$. For this, the most general version of the FLP we will consider, we assume that facility j has an available capacity equal to k_j. The supplier wishes to maximize its net profit, equal to the total revenue from customers less fixed facility operating costs and variable production-to-customer supply costs. We can formulate this profit-maximizing FLP, denoted as FLP1, as follows.

$$\textbf{[FLP1]} \quad \text{Maximize} \quad \sum_{i\in I} r_i z_i - \sum_{j\in J} f_j y_j - \sum_{i\in I}\sum_{j\in J} c_{ij} x_{ij} \tag{8.1}$$

$$\text{Subject to:} \quad \sum_{j\in J} x_{ij} = z_i, \quad i \in I, \tag{8.2}$$

$$\sum_{i\in I} x_{ij} \le k_j y_j, \quad j \in J, \tag{8.3}$$

$$0 \le z_i \le d_i, \quad i \in I, \tag{8.4}$$

$$x_{ij} \ge 0, \quad i \in I, j \in J, \tag{8.5}$$

$$y_j \in \{0,1\}, \quad j \in J. \tag{8.6}$$

Constraints (8.2) and (8.4) together permit the supplier to choose the amount of customer i demand it will satisfy (up to d_i for each $i \in I$). Constraint set (8.3) limits the amount of flow out of facility $j \in J$ to zero if $y_j = 0$, or to the facility's capacity k_j if $y_j = 1$. Constraint set (8.2) permits substituting each z_i variable out of the formulation, resulting in the following reformulation of the problem.

$$\textbf{[FLP1}'\textbf{]} \quad \text{Maximize} \quad \sum_{i\in I}\sum_{j\in J} (r_i - c_{ij}) x_{ij} - \sum_{j\in J} f_j y_j \tag{8.7}$$

$$\text{Subject to:} \quad \sum_{i\in I} x_{ij} \le k_j y_j, \quad j \in J, \tag{8.8}$$

$$0 \le \sum_{j\in J} x_{ij} \le d_i, \quad i \in I, \tag{8.9}$$

$$x_{ij} \ge 0, \quad i \in I, j \in J, \tag{8.10}$$

$$y_j \in \{0,1\}, \quad j \in J. \tag{8.11}$$

Observe that the coefficient of x_{ij} in the objective function (8.7) equals the profit margin for sending a unit from supply point $j \in J$ to customer $i \in I$.

Next consider the single-facility version of FLP1$'$, where $|J| = 1$, which we formulate below, after suppressing the facility index, and letting $\hat{r}_i = r_i - c_{i1}$.

$$\text{Maximize} \quad \sum_{i\in I} \hat{r}_i x_i - fy \tag{8.12}$$

$$\text{Subject to:} \quad \sum_{i\in I} x_i \le ky, \tag{8.13}$$

$$0 \le x_i \le d_i, \quad i \in I, \tag{8.14}$$

$$y \in \{0,1\}. \tag{8.15}$$

Note that we will have either $y = 1$ or $y = 0$. In the latter case, the objective function value equals zero. In the former case, determining the objective function value requires solving a continuous knapsack problem, demonstrating that FLP1 generalizes the continuous knapsack problem.

We next consider a restriction of FLP1 such that if the supplier chooses to satisfy any portion of customer i demand, it must satisfy all d_i units of demand. This *all-or-nothing* customer demand version of the problem may be obtained by replacing each variable z_i in the FLP1 formulation with $z_i' = z_i/d_i$ for each $i \in I$ and requiring each of the z_i' variables to take a value of zero or one. In reformulating this problem it is also convenient to redefine the variable x_{ij} as the percentage of customer i demand supplied by facility j. The resulting reformulation of FLP1, which we denote as FLP1(a) is written as follows.

[FLP1(a)]

$$\text{Maximize} \quad \sum_{i \in I} r_i d_i z_i' - \sum_{j \in J} f_j y_j - \sum_{i \in I} \sum_{j \in J} c_{ij} d_i x_{ij} \tag{8.16}$$

$$\text{Subject to:} \quad \sum_{j \in J} x_{ij} = z_i', \qquad i \in I, \tag{8.17}$$

$$\sum_{i \in I} d_i x_{ij} \leq k_j y_j, \qquad j \in J, \tag{8.18}$$

$$x_{ij} \geq 0, \qquad i \in I, j \in J, \tag{8.19}$$

$$y_j \in \{0, 1\}, \qquad j \in J, \tag{8.20}$$

$$z_i' \in \{0, 1\}, \qquad i \in I. \tag{8.21}$$

This *all-or-nothing* customer demand version of the problem allows splitting a customer's demand among multiple supply points (i.e., for a given demand point i we may have $x_{ij} > 0$ for more than one supply point j). It is often desirable in practice to implement a *single-sourcing* policy in which each demand point receives all supply from a single supply point. Such a requirement may be modeled by requiring each x_{ij} variable in the FLP1(a) formulation to take a value of zero or one. In this case, the z_i' variables may be substituted out of the formulation, resulting in the single-sourcing version of the problem with all-or-nothing demand satisfaction, which may be formulated as follows.

[FLP1(b)]

$$\text{Maximize} \quad \sum_{i \in I} \sum_{j \in J} (r_i - c_{ij}) d_i x_{ij} - \sum_{j \in J} f_j y_j \tag{8.22}$$

$$\text{Subject to:} \quad \sum_{i \in I} d_i x_{ij} \leq k_j y_j, \qquad j \in J, \tag{8.23}$$

$$y_j \in \{0, 1\}, \qquad j \in J, \tag{8.24}$$

$$x_{ij} \in \{0, 1\}, \qquad i \in I, j \in J. \tag{8.25}$$

The single-facility version of this problem variant is equivalent to a 0-1 knapsack problem, which implies that facility location problems with single-sourcing requirements generalize 0-1 knapsack problems (and are thus, in general, $\mathcal{NP}$-Hard optimization problems).

8.2.1 Cost-minimizing version of the FLP

Under suitable conditions on the problem parameter values, an optimal solution will exist that fully satisfies all customer demands. For example, if the minimum value of r_i among all $i \in I$ exceeds the maximum f_j among all $j \in J$ plus the maximum value of $c_{ij}d_i$ among all $i \in I$ and $j \in J$, then this is a sufficient (albeit strong) condition that ensures that each customer's demand is profitable. If, in addition, sufficient capacity exists to accommodate all customer demands, then an optimal solution will exist such that all customer demands are selected (i.e., $z_i = 1$ for all $i \in I$). In such cases, the supplier's total revenue will be fixed, and minimizing the total cost incurred while meeting all customer demands will be equivalent to maximizing net profit. The resulting problem conforms to the more common definition of the FLP in the literature, which we formulate as follows.

[FLP]

$$\text{Minimize} \sum_{j \in J} f_j y_j + \sum_{i \in I} \sum_{j \in J} c_{ij} x_{ij} \tag{8.26}$$

$$\text{Subject to:} \quad \sum_{j \in J} x_{ij} = d_i, \quad i \in I, \tag{8.27}$$

$$\sum_{i \in I} x_{ij} \leq k_j y_j, \quad j \in J, \tag{8.28}$$

$$x_{ij} \geq 0, \quad i \in I, j \in J, \tag{8.29}$$

$$y_j \in \{0, 1\}, \quad j \in J. \tag{8.30}$$

In the above FLP formulation, observe that we have returned to the original definition of x_{ij} as the number of units that flow from supply point $j \in J$ to demand point $i \in I$.

8.2.2 Relationship of the FLP to lot sizing problems

Our discussion of the FLP thus far has not considered an explicit time dimension, while the UELSP and CELSP in Chapters 5 and 6 have an explicit time dimension. Suppose, however, that we view each element of I and J, respectively, as a demand time period and a supply time period, and that each of these sets has cardinality $|T|$. That is, d_i denotes the demand in time period i for $i = 1, \ldots, T$. Similarly, f_j denotes the fixed cost for production in period j for $j = 1, \ldots, T$, while c_{ij} denotes the variable production plus holding cost associated with producing a unit in period j that is used to satisfy demand in period i for $i \geq j$. In terms of the notation from Chapters 5 and 6, and with a slight abuse of notation, we have $c_{ij} = c_j + \sum_{t=j}^{i-1} h_t$ for all $j = 1, \ldots, T$ and $i \geq j$, where c_j is the cost to produce a unit in period j, and h_t is the unit holding cost in period t. In the absence of backordering, we may set $c_{ij} = \infty$ for all $i < j$. It should thus be clear that the FLP formulation in the previous subsection generalizes the CELSP in Chapter 6, which, in turn, generalizes the UELSP from Chapter 5. We can easily account for backordering in this approach by setting $c_{ij} = c_j + \sum_{t=i}^{j-1} b_t$ for $i < j$ instead of ∞, where b_t is the unit backorder cost in period t. As noted in Section 5.3, the tight reformulation of the UELSP corresponds to this facility location formulation approach. The facility location formulation approach for the CELSP also provides a tighter formulation than the initial formulation given in Section 6.2, in the sense that all feasible solutions under the facility location formulation are feasible for the CELSP formulation in Section 6.2, whereas this result does not hold in the opposite direction (see Exercise 6.1).

8.2.3 Single-sourcing version of the FLP and the GAP

We discussed the implications of a single-sourcing requirement for the profit-maximizing version of the FLP earlier in this section. We next formulate the cost-minimizing version of the FLP with single-sourcing requirements, again redefining x_{ij} as the percentage of customer i demand that is satisfied by facility j for all $i \in I$ and $j \in J$. Noting that under single-sourcing requirements each of these x_{ij} variables must be binary, this formulation, denoted as FLP(SS), is written as follows.

[FLP(SS)]

$$\text{Minimize} \sum_{j \in J} f_j y_j + \sum_{i \in I} \sum_{j \in J} c_{ij} d_i x_{ij} \tag{8.31}$$

$$\text{Subject to:} \quad \sum_{j \in J} x_{ij} = 1, \qquad i \in I, \tag{8.32}$$

$$\sum_{i \in I} d_i x_{ij} \leq k_j y_j, \qquad j \in J, \tag{8.33}$$

$$x_{ij} \in \{0, 1\}, \qquad i \in I, j \in J, \tag{8.34}$$

$$y_j \in \{0, 1\}, \qquad j \in J. \tag{8.35}$$

Observe that the special case of FLP(SS) in which $f_j = 0$ for all $j \in J$ corresponds to a generalized assignment problem (GAP) in which the resource consumption of item i is independent of the resource j to which it is assigned. We can, therefore, consider a more general version of FLP(SS) in which customer i demand, when satisfied by facility j, is equal to d_{ij}. This version will therefore have the GAP formulation in Chapter 4 as a special case when all fixed costs equal zero. In practice, however, because a customer's demand is independent of the source facility that meets the demand, this version of the model is not common in the literature.

8.2.4 Set covering and FLP complexity

It is straightforward to show that the single-sourcing requirement is superfluous for the uncapacitated version of the FLP. To see this, observe that if we are given any set of open facilities, then an optimal assignment of customers exists for these facilities such that each customer's demand is fully assigned to the lowest cost open facility for that customer (with ties broken arbitrarily). Because no limit exists on facility capacities, it is feasible to do this, and because an optimal solution contains *some* set of open facilities, the result is implied. For the uncapacitated version of the FLP it is not immediately clear, however, whether this problem is solvable in polynomial time. The answer, unfortunately, is no, unless $\mathcal{P} = \mathcal{NP}$, as we discuss shortly.

Both the uncapacitated FLP and the FLP with single-sourcing requirements may be formulated as set covering or set partitioning problems. The assignment of customer demands to a facility corresponds to a set, with a set's cost equal to the facility's fixed cost plus the associated variable customer assignment costs. Because the objective seeks to minimize the sum of fixed facility and variable customer-to-facility assignment costs, the set covering and set partitioning formulations of the FLP will produce the same optimal solution(s), where the former is a relaxation of the latter. It is also worth noting that the special case of the FLP with single-sourcing requirements in which all capacities are equal, all variable supply costs (c_{ij} values) are zero, and all fixed facility costs are the same is identical to the bin packing problem discussed in Section 3.3.

Cornuejols, Nemhauser, and Wolsey [32] showed that the uncapacitated version of the FLP is $\mathcal{NP}$-Hard via a reduction from a special case of the set covering problem called the vertex covering problem. The vertex covering problem is a combinatorial network optimization problem defined on a graph $G = (V, E)$, where V denotes the vertex set, and E is the edge set. A vertex $v \in V$ *covers* an edge $e \in E$ if and only if e is connected to v. We wish to determine whether a set of k vertices exists (for some positive integer k) that covers all edges in the graph. This vertex covering problem is known to be $\mathcal{NP}$-Complete (see [54]). The reduction provided in [32] works as follows. Given an instance of the vertex covering problem, for each edge we create a corresponding customer, and for each vertex we create a corresponding facility. Let $\tilde{v}(e)$ denote the two nodes incident to edge e in the graph, set the fixed cost of each facility equal to one, and set $c_{e,\tilde{v}(e)} = 0$ for each $e \in E$ and $v \in \tilde{v}(e)$ (if $v \notin \tilde{v}(e)$ then set $c_{e,\tilde{v}(e)} = \infty$). This transformation may be done in polynomial time, implying that if we can solve the uncapacitated FLP in polynomial time, we can also solve the vertex covering problem in polynomial time (which can only occur if $\mathcal{P} = \mathcal{NP}$). Because the uncapacitated FLP is a special case of the FLP with finite capacities, the latter problem class is at least as hard as the former, implying $\mathcal{NP}$-Hardness of the general class of location problems we consider in this chapter. The following section discusses a dual-based method for solving the uncapacitated FLP.

8.3 Dual-Ascent Method for the Uncapacitated FLP

One of the most successful approaches for solving the uncapacitated version of the FLP, which we denote as the UFLP, was provided by Erlenkotter [42]. This solution approach begins with the following "tight" formulation of the

UFLP.

[UFLP]

$$\text{Minimize} \quad \sum_{j \in J} f_j y_j + \sum_{i \in I} \sum_{j \in J} \hat{c}_{ij} x_{ij} \tag{8.36}$$

$$\text{Subject to:} \quad \sum_{j \in J} x_{ij} = 1, \qquad i \in I, \tag{8.37}$$

$$x_{ij} \leq y_j, \qquad i \in I, j \in J, \tag{8.38}$$

$$x_{ij} \geq 0, \qquad i \in I, j \in J, \tag{8.39}$$

$$y_j \in \{0, 1\}, \qquad j \in J. \tag{8.40}$$

In this formulation, we have used the notation $\hat{c}_{ij} = c_{ij} d_i$ for all $i \in I$ and $j \in J$, where $\hat{c}_{ij}$ corresponds to the total variable cost of assigning customer i to facility j. Observe that constraint set (8.38) corresponds to a *disaggregated* version of the following constraint set:

$$\sum_{i \in I} x_{ij} \leq |I| y_j, \quad j \in J. \tag{8.41}$$

Despite requiring more constraints in total, constraint set (8.38) leads to a tighter formulation than with the use of (8.41), as the latter constraint set admits solutions to the corresponding LP relaxation (obtained by replacing (8.40) with $0 \leq y_j \leq 1, j \in J$) which the former constraint set does not. To see this, suppose that $y_{j'} = 0.25$ for some $j' \in J$, while $|I| = 12$. Then the right-hand side of (8.41) will equal 3 for $j = j'$, which allows $x_{ij'} = 1$ for one or more (i, j') pairs. Constraint set (8.38), on the other hand, does not permit any $x_{ij'}$ value to exceed 0.25, while ensuring that (8.41) is also satisfied.

We next consider the LP relaxation of the UFLP, with (8.40) replaced by

$$y_j \geq 0, \quad j \in J, \tag{8.42}$$

and observe that we need not explicitly include the constraint set $y_j \leq 1$, $j \in J$, because these variables will be as small as possible in an optimal solution, and no x_{ij} value can be greater than one as a result of constraints (8.37) and (8.39). Let v_i, $i \in I$ denote a set of dual variables associated with constraint set (8.37); similarly, let w_{ij}, $i \in I$, $j \in J$ denote a set of dual variables corresponding to constraint set (8.38). The dual of the LP relaxation of the UFLP can then be formulated as follows.

[UFLP(D1)]

$$\text{Maximize} \quad \sum_{i \in I} v_i \tag{8.43}$$

$$\text{Subject to:} \quad \sum_{i \in I} w_{ij} \leq f_j, \qquad j \in J, \tag{8.44}$$

$$v_i - w_{ij} \leq \hat{c}_{ij}, \qquad i \in I, j \in J, \tag{8.45}$$

$$w_{ij} \geq 0, \qquad i \in I, j \in J. \tag{8.46}$$

Because each w_{ij} variable is nonnegative and is bounded from below by $v_i - \hat{c}_{ij}$, and because we may set w_{ij} to its smallest possible value without affecting the objective function value, we can make the substitution $w_{ij} = \max\{v_i - \hat{c}_{ij}, 0\}$ without loss of optimality. This permits formulating the dual problem much more compactly as follows.

[UFLP(D2)]

$$\text{Maximize} \quad \sum_{i \in I} v_i \tag{8.47}$$

$$\text{Subject to:} \quad \sum_{i \in I} \max\{v_i - \hat{c}_{ij}, 0\} \leq f_j, \quad j \in J. \tag{8.48}$$

The dual-ascent approach begins with a dual feasible solution and successively increases dual variable values as feasibility permits. Because the dual objective function value provides a lower bound on the LP relaxation solution value, solving the dual problem provides a lower bound on the optimal solution of the UFLP. The dual ascent approach also lends itself to heuristic methods for providing feasible solutions for the UFLP, as we will show later. The UFLP(D2) formulation has a special structure, with an objective function that seeks to maximize the unweighted sum of a single set of dual variables, and a constraint set that limits the total increase of these variables. An initial dual feasible solution (with nonzero objective value) may be easily obtained as follows. For each customer $i \in I$, identify the facility with the smallest variable cost for supplying customer i, let $j_1(i)$ denote this facility, and set $v_i = \hat{c}_{ij_1(i)}$. Clearly, this solution is feasible for UFLP(D2), as it consumes none of the capacity of the constraint set (8.48). It should also be clear that this solution does not solve the dual problem, as feasible solutions exist with higher values of the v_i variables which do not violate any constraints (assuming $f_j > 0$ for all $j \in J$).

To begin the dual-ascent approach, for each customer i, we sort values of $\hat{c}_{ij}$ in nondecreasing order, letting $p(i)$ denote the pth facility in this sorted order for customer i, and letting $c_i^{p_i}$ denote the corresponding value of $\hat{c}_{ij}$. At each iteration, the dual-ascent approach will attempt to increase v_i for some value of i to the next higher value of $c_i^{p_i}$. That is, initially, $v_i = c_i^1$, and when we first consider increasing v_i at some step of the algorithm, we will attempt to increase v_i to c_i^2 if possible. For completeness of the algorithm, we add a dummy facility $\tilde{j}$ such that $\hat{c}_{i\tilde{j}} = \infty$ for each $i \in I$ (although we do not add an additional dual constraint corresponding to this facility).

Let s_j denote the slack associated with the constraint (8.48) for facility j, i.e.,

$$s_j = f_j - \sum_{i \in I} \max\{v_i - \hat{c}_{ij}, 0\}, \quad j \in J. \tag{8.49}$$

Under the initial solution we described previously, $s_j = f_j$ for each $j \in J$.

The dual-ascent procedure is formalized in Algorithm 8.1, which works as follows. At any step of the algorithm in which we consider changing the value of v_i for any i, we have $v_i = c_i^{p_i}$ for some value of p_i (initially $p_i = 1$ for each i).

Beginning with $i = 1$, we set $\Delta_i = \min_{j \in J}\{s_j : v_i - \hat{c}_{ij} \geq 0\}$. This initial value of Δ_i provides an upper bound on the amount we may increase v_i without violating any constraint. We compare this value of Δ_i with $c_i^{p_i+1} - v_i$, as we would like to increase v_i to $c_i^{p_i+1}$ if possible. If Δ_i is greater than $c_i^{p_i+1} - v_i$, then we will set $\Delta_i = c_i^{p_i+1} - v_i$, implying that we will increase v_i to $c_i^{p_i+1}$. If this is the case, then we set $p_i \leftarrow p_i + 1$ and set an indicator variable, δ, equal to one to indicate that we have not yet exhausted all available capacity for increasing the value of v_i. Otherwise δ remains at zero and the increase in v_i by Δ_i exhausts the capacity of some constraint. In addition to increasing v_i to $v_i + \Delta_i$, we decrease s_j by Δ_i for each j such that $v_i \geq \hat{c}_{ij}$. After considering item i, we then move on to item $i + 1$ and repeat the process unless $i = |I|$. If $i = |I|$ and $\delta = 1$, then available capacity remains, and we restart the procedure at $i = 1$; otherwise, we terminate the procedure.

Algorithm 8.1 Erlenkotter's dual-ascent algorithm (adapted from [42]).

1: Begin with initial solution that sets $v_i = c_i^1$ for all $i \in I$; let $s_j = f_j$ for each $j \in J$, and set $p_i = 1$ for all $i \in I$.
2: Set $i = 1$ and $\delta = 0$.
3: Let $\Delta_i = \min\{\min\{s_j : v_i \geq \hat{c}_{ij}\}, c_i^{p_i+1} - v_i\}$; if the latter argument gives the minimum, set $\delta = 1$ and $p_i \leftarrow p_i + 1$.
4: For each $j \in J$ such that $v_i \geq \hat{c}_{ij}$, decrease s_j by Δ_i; increase v_i by Δ_i.
5: If $i < |I|$, set $i \leftarrow i + 1$, and go to Step 3.
6: If $\delta = 1$, go to Step 2; otherwise, end.

The algorithm terminates with a feasible dual solution for which some subset of the constraint set's slack variables will equal zero. The following comprise the complementary slackness conditions, which are necessary and sufficient for validating the optimality of a solution to the LP relaxation:

$$y_j s_j = 0, \qquad j \in J, \tag{8.50}$$

$$(y_j - x_{ij}) \max\{v_i - \hat{c}_{ij}, 0\} = 0, \qquad i \in I, j \in J. \tag{8.51}$$

Using the feasible dual solution and these complementary slackness conditions, Erlenkotter [42] suggests creating a primal solution that is feasible for UFLP (i.e., the y_j variables are binary) and attempts to satisfy (8.50) and (8.51) as closely as possible. To do this, let J^* denote the set of facilities such that $s_j = 0$ in the corresponding dual solution, and for some subset J^+ of J^* (we will discuss the composition of this subset shortly), we first set $y_j = 1$ for $j \in J^+ \subseteq J^*$ and $y_j = 0$ for $j \notin J^+$, which ensures that (8.50) is satisfied. Next, for each $i \in I$, let $j^+(i)$ denote the facility in J^+ with the minimum value of $\hat{c}_{ij}$ and set $x_{ij^+(i)} = 1$ (for all other $j \in J$, set $x_{ij} = 0$). If the resulting solution satisfies (8.51), then this solution is optimal for both the LP relaxation and UFLP. This is not, however, guaranteed, and the resulting (heuristic) solution may fail to satisfy (8.51). These violations occur when

$v_i > \hat{c}_{ij}$ for some $j \in J^+$ with $j \neq j^+(i)$ (for such (i,j) pairs, $y_j = 1$, $x_{ij} = 0$, and $v_i > \hat{c}_{ij}$, implying that the left-hand side of (8.51) is nonzero).

Observe that we defined J^+ as a subset of J^*, the set of all facilities such that $s_j = 0$. Erlenkotter [42] suggests a method of approximating a minimal set J^+ based on the properties of the dual solution. Note that for each $i \in I$ we require some facility in J^+ such that $v_i \geq \hat{c}_{ij}$. If for some $i \in I$ there is only one facility $j \in J^*$ such that $v_i \geq \hat{c}_{ij}$, then such a facility is deemed essential in creating a primal solution. We can thus begin with $J^+ = J^*$ and sequentially remove non-essential facilities from the solution (i.e., if a facility $j' \in J^*$ is in J^+ and no customer i exists such that facility j' is the only facility in J^+ such that $v_i \geq \hat{c}_{ij}$, then such a facility may be deemed as non-essential).

Next observe that the solution obtained via Algorithm 8.1 depends on the sequence in which customer demands in the set I are processed. Erlenkotter [42] experimented with different sequencing strategies, although no strategy appears to be uniformly superior. He also noted that it is often possible to adjust and improve the solution obtained by Algorithm 8.1 via a dual-adjustment procedure. This adjustment procedure considers those $i \in I$ such that (8.51) is violated. For such an $i' \in I$, reducing the value of $v_{i'}$ (to the next lower value of $\hat{c}_{i'j}$) will lead to a positive value of s_j for two or more constraints (recall that (8.51) is violated when $v_{i'} > \hat{c}_{i'j}$ for more than one $j \in J^+$, and j can only be in J^+ if $s_j = 0$). In order to absorb this slack created by the reduction of $v_{i'}$, we attempt to increase the values of v_i for $i \neq i'$ via the dual-ascent algorithm. If two or more v_i values can increase to absorb the slack as a result of the reduction in $v_{i'}$, then the dual objective will increase; if only one v_i can increase, then the dual objective will remain the same. Note that it is possible that the dual-ascent algorithm will leave the slack on one or more constraints unabsorbed, in which case the number of elements of the set J^+ must decrease. We refer the reader to the paper by Erlenkotter [42] for details on the dual-adjustment procedure, as well as the results of numerical tests that demonstrate the effectiveness of the dual-ascent and dual-adjustment algorithm in providing high-quality solutions for the UFLP.

Example 8.1 We consider an example problem with four facilities and six customers, based on a class of problems identified in [62]. In this example, each facility has a fixed cost of 10, and each customer has a demand of 10. The $\hat{c}_{ij}$ values are shown in the table below.

	Customer					
Facility	1	2	3	4	5	6
1	10	10	10	30	30	30
2	10	30	30	10	10	30
3	30	10	30	10	30	10
4	30	30	10	30	10	10

The left-hand side of each of the four dual constraints contains a sum of three terms of the form $\max\{v_i - 10, 0\}$ and three terms of the form $\max\{v_i -$

$30, 0\}$; this summation must be less than or equal to 10. For example, the dual constraint associated with facility 1 is written as

$$\sum_{i=1}^{3} \max\{v_i - 10, 0\} + \sum_{i=4}^{6} \max\{v_i - 30, 0\} \leq 10.$$

The dual ascent procedure begins with $v_i = 10$ for $i = 1, \ldots, 6$. Suppose we start by increasing v_1, which can be increased by 10 (to a value of 20) before absorbing all of the capacity of the first two dual constraints. None of the variables v_2 through v_5 may be further increased without violating at least one capacity limit. The dual variable v_6 can increase to 20, however, at which point it uses all of the capacity of the third and fourth dual constraints. The resulting solution maximizes the dual with an objective function of 80 (which equals the LP relaxation optimal solution value). Observe that if we begin by increasing v_2 (v_3) first, then the resulting solution will increase v_2 and v_5 (v_3 and v_4) to 20, with all other variables equal to 10. The resulting solution is an alternative dual optimal solution.

We next use the dual solution to create a primal feasible solution. The set J^*, i.e., the facilities such that $s_j = 0$, consists of all facilities. Initially, none of these facilities is deemed as essential (each customer has $v_i \geq \hat{c}_{ij}$ for two facilities). Thus, we can close any one of the facilities in order to obtain a minimal set J^+. Suppose we close facility 1. Then facility 2 is essential for customer 1, facility 3 is essential for customer 2, and facility 4 is essential for customer 3. With facilities 2, 3, and 4 open, we set $x_{12} = x_{23} = x_{34} = x_{42} = x_{52} = x_{63} = 1$, and all other x_{ij} variables equal to 0. The resulting solution has fixed costs equal to 30 and customer assignment costs equal to 60, which, as it turns out, is an optimal solution for this instance of the UFLP. □

8.4 Approximation Algorithms for the Metric UFLP

The class of *metric* facility location problems assumes that the facility assignment cost ($\hat{c}_{ij}$) values are symmetric and are proportional to the Euclidean distances between customers and facilities (because of this we will use the terms "distance" and "cost" interchangeably). In the metric location problem class, these distances, as well as all customer-to-customer and facility-to-facility distances, must obey the triangle inequality. This implies that for any pair of facilities (j_1, j_2) and any pair of customers (i_1, i_2), the condition $c_{i_1 j_1} \leq c_{i_1 j_2} + c_{i_2 j_1} + c_{i_2 j_2}$ must hold. Other than this assumption on distances (and the associated distance-based costs), the uncapacitated version of this problem is identical to the UFLP.

In Chapters 2 and 6, we discussed approximation algorithms for the knapsack and capacitated lot sizing problems, respectively. These fully polynomial time approximation schemes required defining an optimality tolerance ϵ, and led to running times that were polynomial in the size of the input data and $1/\epsilon$. In contrast, an algorithm that gives a *constant-factor approximation* runs in polynomial time in the size of the input data and provides a solution with cost that is no more than $1+\gamma$ times the optimal solution value for some fixed and nonnegative number γ. A great deal of work in the past twenty years has focused on developing constant-factor approximation algorithms, in particular for the metric UFLP, which we denote as MUFLP. Our goal in this section is to give the reader an idea of how such constant-factor approximation algorithms work, as well as to summarize the best known results to date from the literature. To do this, we first review in some detail the approach taken to obtain the first-known constant-factor approximation algorithm for the MUFLP, provided by Shmoys, Tardos, and Aardal [106].

The approach of Shmoys et al. [106] begins with a solution to the LP relaxation of the MUFLP (constant-factor approximation approaches often use an LP relaxation solution or a greedy algorithm as a starting point). They define the concept of g-closeness of an LP relaxation solution as follows. Given $g_i \geq 0$ for each $i \in I$, an LP relaxation solution is g-close if, for each $x_{ij} > 0$, the corresponding value of $\hat{c}_{ij}$ is less than or equal to g_i (thus, given an LP relaxation solution, for each $i \in I$, we may set g_i equal to the maximum distance from customer i to any facility j with positive flow to customer i, and claim that the resulting solution is g-close). Their first important result says that if we are given a g-close solution (x', y') for the LP relaxation, then it is possible to find, in polynomial time, an integer-feasible $3g$-close solution (x'', y'') that also satisfies $\sum_{j \in J} f_j y_j'' \leq \sum_{j \in J} f_j y_j'$.

Demonstrating this result requires an algorithm that begins with the LP relaxation solution (x', y') and finds the integer-feasible solution (x'', y''). The algorithm may be described as follows. We start with an LP relaxation solution (x', y'), and we apply an approach that clusters groups of demand points around a center, located at some demand point. This clustering begins with the demand point i with the smallest value of g_i (which is the maximum distance from any supply point with positive flow to the demand point in the solution (x', y')); call this demand point i_0. For each demand point i, define the neighborhood of the demand point, $\mathcal{N}(i)$, as the set of all facilities with positive flow to the demand point. Beginning with i_0, we assign to this cluster all demand points that receive positive flow from facilities in the neighborhood $\mathcal{N}(i_0)$. Next, among all unassigned demand points, we repeat this process until all demand points have been assigned to clusters. The output of this process is a set of demand-point clusters, each with a cluster center. As a result of the algorithm's construction, note that the neighborhoods of the cluster centers are disjoint.

Next, for each cluster, we open a facility; in particular, given a cluster center i_0, for example, we open the facility in $\mathcal{N}(i_0)$ with the smallest fixed cost. All customers in the same cluster are assigned to the same facility; let f_{i_0} denote the fixed cost of the facility selected for cluster center i_0. Next, observe that the LP relaxation solution and the definition of the neighborhood of a demand point require $\sum_{j \in \mathcal{N}(i)} x'_{ij} = 1$; this, in turn, implies that $\sum_{j \in \mathcal{N}(i)} y'_j \geq 1$ as a result of constraint set (8.38). Because we have selected the facility with the lowest fixed cost for each cluster center, we must therefore have $f_{i_0} \leq \sum_{j \in \mathcal{N}(i_0)} f_j y'_j$ for a cluster centered at i_0. This implies that the solution satisfies $\sum_{j \in J} f_j y''_j \leq \sum_{j \in J} f_j y'_j$.

Shmoys et al. [106] next provide a bound on the distance between any customer and its assigned facility. Consider a demand point k and suppose it is assigned to a cluster centered at i_0 (see Figure 8.1). This can only occur if demand points k and i_0 have at least one common neighbor facility. The distance from customer k to this common neighbor facility is no more than g_k, while the distance from customer i_0 to this common neighbor facility is no more than g_{i_0}. In addition, the distance from customer i_0 to the facility assigned to the cluster is no more than g_{i_0}, because this facility must be in the neighborhood of i_0. Thus, the distance from customer k to its assigned facility is no more than $2g_{i_0} + g_k$. Because i_0 was selected as the cluster center (and not k), this implies $g_{i_0} \leq g_k$, which further implies that the distance from customer k to its assigned facility is no more than $3g_k$. This is certainly true for any customer i, which implies that the total assignment costs associated with the integer feasible solution constructed do not exceed $\sum_{i \in I} 3g_i$.

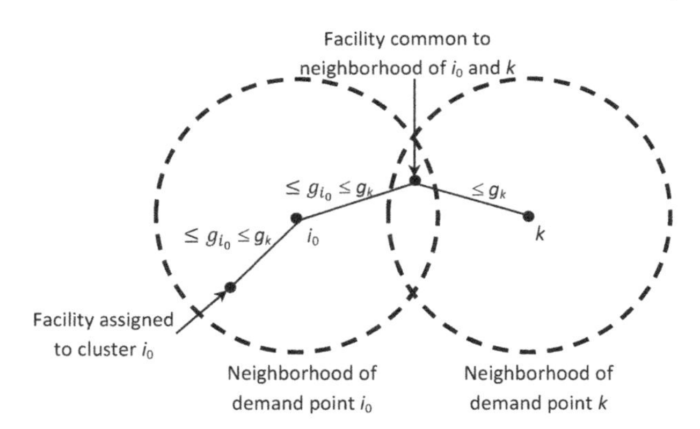

FIGURE 8.1
Illustration of bound on distance from demand point to assigned facility.

Next, consider the LP relaxation solution (x', y') and suppose that the complementary optimal dual solution to UFLP(D1) is given by (v', w'). Let z^*_{LP} denote the optimal solution value for the LP relaxation and note that $z^*_{LP} = \sum_{i \in I} v'_i$. Moreover, if $x'_{ij} > 0$, then $v'_i = \hat{c}_{ij} + w'_{ij} \geq \hat{c}_{ij}$ (by complementary slackness and the nonnegativity of the w_{ij} variables). By the definition of g_i (g_i equals the maximum $\hat{c}_{ij}$ such that $x'_{ij} > 0$), we thus have $v'_i \geq g_i$, which implies that the assignment costs associated with the constructed integer feasible solution are not greater than $\sum_{i \in I} 3g_i \leq \sum_{i \in I} 3v'_i = 3z^*_{LP}$.

We also know that the fixed facility costs associated with the constructed integer feasible solution satisfy $\sum_{j\in J} f_j y''_j \leq \sum_{j\in J} f_j y'_j \leq z^*_{LP}$. As a result, the total cost of the integer feasible solution is bounded by $4z^*_{LP}$, implying a constant-factor approximation bound of 4 for this solution. Observe that the bound on assignment costs is much weaker than that of the fixed facility costs. This results because the algorithm always chooses the cheapest facility in each cluster center neighborhood, perhaps at the expense of high assignment costs. Although the best known constant-factor approximations today are much better than this, this was the first such result for the MUFLP.

8.4.1 Randomization and derandomization

Chudak and Shmoys [27] substantially improved the clustering algorithm via randomization and derandomization techniques. We next summarize these approaches and results. Their approach first clusters customers as described previously, although it alters the method of opening facilities. They distinguish facilities based on whether they fall within the neighborhood of a cluster center. Those that fall within the neighborhood of a cluster center are called central facilities; the remaining facilities are non-central facilities. They open one central facility in each cluster, as before; however, for a cluster center i_0, and independently for each cluster, their randomized algorithm opens facility j in the neighborhood of i_0 with probability $x'_{i_0 j}$. Non-central facilities are opened with probability y'_j. Each demand is then assigned to the closest open facility. Because they are able to show without loss of generality that the LP relaxation solution is complete, in the sense that $x'_{ij} = y'_j$ if $x'_{ij} > 0$, they are thus able to show that the expected total facility cost associated with this approach equals $\sum_{j\in J} f_j y'_j$. However, under this randomized algorithm, they can also show that the expected assignment cost for any customer k is at most $\sum_{j\in J} \hat{c}_{kj} x'_{kj} + (3/e) v'_k$,[1] which implies that the expected total assignment cost is no more than $\sum_{i\in I}\sum_{j\in J} \hat{c}_{ij} x'_{ij} + (3/e)\sum_{i\in I} v'_i = \sum_{i\in I}\sum_{j\in J} \hat{c}_{ij} x'_{ij} + (3/e) z^*_{LP}$. As a result, the expected cost of the solution of the randomized algorithm is no more than $\sum_{j\in J} f_j y'_j + \sum_{i\in I}\sum_{j\in J} \hat{c}_{ij} x'_{ij} + (3/e) z^*_{LP} = (1 + 3/e) z^*_{LP}$.

Next, Chudak and Shmoys [27] show that a slight change in the way cluster centers are chosen can further improve this bound on the expected objective function value. In obtaining the results discussed thus far, a critical factor was the bound on the expected distance from a cluster center to the facility that serves the cluster center. For this bound, recall that $v'_i \geq g_i$ was used. However,

[1] As Chudak and Shmoys [27] show, the probability that a non-cluster-center customer k has no open facility in its neighborhood equals $\prod_{j\in J}(1 - x'_{kj})$. Because $1 - x \leq e^{-x}$ when $x > 0$, this quantity is no bigger than $\prod_{j\in J: x'_{jk} > 0} e^{-x'_{jk}}$, which equals $e^{-\sum_{j\in J} x'_{kj}}$. Because $\sum_{j\in J} x'_{kj} = 1$, this quantity equals 1/e. Thus, the probability that a non-cluster-center customer k has no open facility in its neighborhood is no more than 1/e.

in the randomized algorithm, facility j is selected to serve cluster center i_0 with probability $x'_{i_0 j}$; as a result, the expected distance from a cluster center to the facility that serves it equals $\sum_{j\in J}\hat{c}_{i_0 j}x'_{i_0 j}$. Thus, instead of choosing cluster centers based on the smallest available g_i values, they choose clusters based on the smallest value of $v'_i + \sum_{j\in J}\hat{c}_{ij}x'_{ij}$. This clustering improvement permits them to bound the expected total assignment cost by $\sum_{i\in I}\sum_{j\in J}\hat{c}_{ij}x'_{ij} + (2/e)z^*_{LP}$, leading to an overall expected value of the solution bound of $(1 + 2/e)z^*_{LP}$.

Chudak and Shmoys [27] then apply a very clever derandomization approach based on conditional expectations in order to determine which facilities should be open, such that the resulting solution value is no greater than the expected solution bound of $(1+2/e)z^*_{LP}$. We describe the essence of this derandomization approach as follows. Let W denote a random variable for the solution value obtained by the randomized algorithm, where $E[W] \le (1+2/e)z^*_{LP}$, and note that for any given instance of MUFLP, $E[W]$ may be computed after solving the LP relaxation. In the randomized algorithm, facility j is opened with some probability $u_j \in [0, 1]$; let U_j denote a random variable associated with this probability for each $j \in J$. A derandomized algorithm requires assigning U_j a value of 0 or 1 for each $j \in J$. Next note that

$$E[W] = E[W|U_j = 1]P\{U_j = 1\} + E[W|U_j = 0]P\{U_j = 0\}, \qquad (8.52)$$

i.e., $E[W]$ is a convex combination of the two conditional probabilities in (8.52). For the non-central facilities, the probabilities of opening the facilities are independent, which implies that we can sequentially compute (8.52) for each non-central facility independently. That is, suppose these facilities are indexed from 1 to n. For facility 1, compute (8.52) and if $E[W|U_1 = 1] \le E[W|U_1 = 0]$ then set $\bar{u}_1 = 1$, implying that facility 1 is open; otherwise, set $\bar{u}_1 = 0$. Then, given values of $\bar{u}_j$ for $j = 1, \ldots, l-1$, compute $E[W|U_1 = \bar{u}_1, U_2 = \bar{u}_2, \ldots, U_{l-1} = \bar{u}_{l-1}, U_l = 1]$ and $E[W|U_1 = \bar{u}_1, U_2 = \bar{u}_2, \ldots, U_{l-1} = \bar{u}_{l-1}, U_l = 0]$. If the former is no greater than the latter, set $\bar{u}_l = 1$; otherwise, set $\bar{u}_l = 0$. This approach is sequentially applied for all n non-central facilities, and it ensures that the conditional expected value of W, given that $U_1 = \bar{u}_1, U_2 = \bar{u}_2, \ldots, U_{n-1} = \bar{u}_{n-1}, U_n = \bar{u}_n$ is less than or equal to $E[W]$. Let $\bar{W}$ denote the conditional value of W given the selection of $\bar{u}_j$ values for the non-central facilities, i.e., $\bar{W} = W|U_1 = \bar{u}_1, U_2 = \bar{u}_2, \ldots, U_{n-1} = \bar{u}_{n-1}, U_n = \bar{u}_n$.

Next, we need to consider the central facilities, which are a bit more complicated because the probabilities associated with opening these facilities are not independent. However, we know that if we do open a central facility for a cluster, then all other neighboring facilities for the cluster center must be closed. As a result, for a cluster centered at i_0, we can compute $E[\bar{W}]$ using

$$E[\bar{W}] = \sum_{j\in\mathcal{N}(i_0)} E[\bar{W}|U_j = 1]P\{U_j = 1\}, \qquad (8.53)$$

where $\mathcal{N}(i_0)$ is the set of central facilities in the neighborhood of cluster center

i_0. We need only compute $E[\bar{W}|U_j = 1]$ for each $j \in \mathcal{N}(i_0)$ and set $U_j = 1$ for the facility that gives the lowest of these conditional expectations (and $U_j = 0$ for all other $j \in \mathcal{N}(i_0)$). This ensures that the resulting conditional expectation is no greater than $E[\bar{W}]$, which is less than or equal to $E[W]$. We can apply this approach sequentially to each cluster, as the probabilities across clusters are independent of each other. This approach will ensure that the resulting solution value is no greater than $E[W]$, which is bounded by $(1 + 2/e)z^*_{LP}$.

Example 8.2 Consider the problem from Example 8.1. The optimal LP relaxation solution value equals 80 and sets $y_j = 0.5$ for all $j = 1, \ldots, 4$, with $x_{11} = x_{21} = x_{31} = x_{12} = x_{42} = x_{52} = x_{23} = x_{43} = x_{63} = x_{34} = x_{54} = x_{64} = 0.5$ (all other variables equal zero). This gives neighborhood definitions of $\mathcal{N}(1) = \{1, 2\}, \mathcal{N}(2) = \{1, 3\}, \mathcal{N}(3) = \{1, 4\}, \mathcal{N}(4) = \{2, 3\}, \mathcal{N}(5) = \{2, 4\}$, and $\mathcal{N}(6) = \{3, 4\}$. Note that $g_i = 10$ for each $i = 1, \ldots, 6$. If we begin with $i_0 = 1$, then because customers 2, 3, 4, and 5 each receive positive flow from facilities 1 and 2 (which are in $\mathcal{N}(i_0)$), we add these customers to the cluster with customer 1. Customer 6 then forms its own cluster. We open a single facility per cluster and, since all $f_j = 10$, we may open either facility 1 or 2 for the first cluster, and either facility 3 or 4 for the second cluster. Suppose we open facilities 1 and 3 for their respective clusters, with customers 1 through 5 assigned to facility 1 and customer 6 assigned to facility 3. The resulting solution value equals 120 (which exceeds the optimal solution value of 90 by 30). Note that it is possible to locally improve this solution by moving customer 4 to facility 3, which reduces the objective by 20 to 100. The randomization/derandomization approach does not, unfortunately, improve the solution value, because all of the facilities are deemed as central facilities, with facilities 1 and 2 associated with the first cluster, and facilities 3 and 4 associated with the second cluster. Because the procedure only opens one central facility per cluster, we will end up with two open facilities only, while an optimal solution for this instance of the UFLP requires three open facilities. □

Recall that the quantity $1 + 2/e < 1.736$ gives the factor for the constant-factor approximation provided by Chudak and Shmoys [27]. Several researchers have subsequently continued to improve this bound. In 2006, Mahdian, Ye, and Zhang [82] used a greedy algorithm along with a cost-scaling approach to reduce the best available constant-factor approximation for the MUFLP to 1.52. They also provided a summary of the history of improvements to the best available constant-factor approximation for this problem up to that time. Byrka and Aardal [21] further improved this bound to 1.5, while Li [79] obtained a bound on the expected solution value of 1.488. It is important to note that Guha and Khuller [62] demonstrated that a bound better than 1.463 is not likely to be possible under a technical condition on the class of problems in $\mathcal{NP}$, a condition which was improved by Sviridenko

[108] to state "unless $\mathcal{P} = \mathcal{NP}$." Thus, the best known results at this time are approaching this theoretical bound.

8.5 Solution Methods for the General FLP

As noted in the formulation of the FLP, the general version of the problem contains finite facility capacities. While the UFLP always has an optimal solution such that each customer's demand is fully allocated to a single facility, this is not necessarily the case for the FLP. Because this is sometimes a desirable property in practice, recall that we considered the FLP with single-sourcing, denoted as FLP(SS), in Section 8.2.3. Because the relaxation of the binary restrictions on the assignment variables (the x_{ij} variables) in FLP(SS) is completely equivalent to the FLP formulation in Section 8.2.1, we will use the FLP(SS) formulation as the basis for our discussion in this section. That is, when the binary restrictions are replaced by the simple constraints,

$$x_{ij} \geq 0, \quad i \in I, j \in J, \tag{8.54}$$

the FLP(SS) and FLP formulations are the same (recall that constraint set (8.32) implies that we do not explicitly require the constraint set $x_{ij} \leq 1$ for all $i \in I$ and $j \in J$ in order for this equivalence to hold). Thus we can use the formulation of FLP(SS) in our discussion of solution methods for the FLP, with the only distinction between the restricted version (with single-sourcing requirements) and the FLP being whether the assignment variables must take binary values or may take any value between (and including) zero and one. The solution methods that have been shown to be effective for solving the FLP have already been discussed throughout this book. In particular, the effective solution methods we will discuss throughout the remainder of this chapter include the use of Lagrangian relaxation, valid inequalities, and approximation algorithms.

8.5.1 Lagrangian relaxation for the FLP

In developing a Lagrangian relaxation strategy, it is important to keep in mind the *integrality property* discussed in Section 4.3. That is, if the feasible region of the Lagrangian relaxation problem (the remaining problem after the relaxation of a constraint set) contains all integer extreme points, then the best bound provided by solving the Lagrangian dual problem will be no better than the bound provided by solving the problem's LP relaxation. As a result, it is immediately clear that if we begin with the FLP(SS) formulation and consider a Lagrangian relaxation in which all constraints of the form (8.33) are relaxed, then the remaining feasible region, even when the binary restrictions are relaxed for both the x_{ij} and y_j variables, contains all binary extreme points (one may verify the total unimodularity of the resulting constraint matrix, or show that any fractional feasible solution can be expressed as a convex combination of binary feasible solutions). On the other hand, if we add the redundant constraint set

$$x_{ij} \leq y_j, \quad i \in I, j \in J, \tag{8.55}$$

as well as the valid constraint set

$$\sum_{j \in J} k_j y_j \geq \sum_{i \in I} d_i, \tag{8.56}$$

then the relaxation of all constraints of the form (8.33) may lead to bounds that are much better than the LP relaxation, via so-called Lagrangian decomposition techniques (see Guignard and Kim [63] and Van Roy [113]).

We next illustrate a straightforward Lagrangian relaxation approach for the FLP for which the integrality property does not hold. This approach was introduced by Nauss [86] for the case in which single sourcing is not required. Let λ_i, $i \in I$, denote a set of nonnegative Lagrangian multipliers associated with the demand assignment constraint set (8.32),[2] and suppose that single sourcing is not required. Recalling that $\hat{c}_{ij} = c_{ij} d_i$, then the resulting Lagrangian relaxation formulation with the addition of the valid constraint (8.56) is written as

[LR(FLP)]

$$\text{Minimize} \quad \sum_{j \in J} f_j y_j + \sum_{i \in I} \sum_{j \in J} (\hat{c}_{ij} - \lambda_i) x_{ij} \tag{8.57}$$

$$\text{Subject to:} \quad \sum_{i \in I} d_i x_{ij} \leq k_j y_j, \quad j \in J, \tag{8.58}$$

$$\textstyle\sum_{j \in J} k_j y_j \geq \sum_{i \in I} d_i, \tag{8.59}$$

$$x_{ij} \geq 0, \quad i \in I, j \in J, \tag{8.60}$$

$$y_j \in \{0, 1\}, \quad j \in J. \tag{8.61}$$

Note that we have omitted the constant term $\sum_{i \in I} \lambda_i$ from the Lagrangian objective above. For a given vector λ of Lagrangian multiplier values and in the absence of the valid inequality (8.59), the Lagrangian relaxation problem decomposes into a set of $|J|$ continuous knapsack problems. Nauss [86] made the clever observation that if for some $j = j'$, we have $y_{j'} = 1$, then we can solve the associated continuous knapsack problem for facility j' to determine the resulting optimal values of the $x_{ij'}$ variables. Letting $x^*_{ij'}(\lambda)$ denote the optimal value of $x_{ij'}$ when $y_{j'} = 1$ and for the given vector λ, we can compute the optimal *value* of the knapsack problem associated with facility j', denoted as $v(\lambda, j')$, as

$$v(\lambda, j') = f_{j'} + \sum_{i \in I} (\hat{c}_{ij'} - \lambda_i) x^*_{ij'}(\lambda), \tag{8.62}$$

[2] Although (8.32) is an equality constraint set, we can replace this equality with a greater-than-or-equal-to inequality in the FLP(SS) formulation without loss of optimality, as the optimal solution is the same in either case, as a result of the minimization objective. We can therefore confine our attention to λ_i values that are nonnegative.

for any $j' \in J$. As a result, solving LR(FLP) is equivalent to solving the following 0-1 knapsack problem:

$$\text{Minimize} \quad \sum_{j \in J} v(\lambda, j) y_j \tag{8.63}$$

$$\text{Subject to:} \quad \sum_{j \in J} k_j y_j \geq \sum_{i \in I} d_i, \tag{8.64}$$

$$y_j \in \{0, 1\}, \qquad j \in J. \tag{8.65}$$

Therefore, for a given vector λ of Lagrangian multipliers, the Lagrangian relaxation problem may be solved by first solving $|J|$ continuous knapsack problems, and then solving a single 0-1 knapsack problem. It is easy to see how a single-sourcing requirement may be accommodated using this same Lagrangian relaxation approach. Instead of solving a continuous knapsack problem for each $j \in J$ to determine each corresponding $x^*_{ij}(\lambda)$ value, we need to solve a 0-1 knapsack problem, resulting in the need to solve $|J| + 1$ instances of the 0-1 knapsack problem for any choice of λ. Letting $Z_{LR}(\lambda)$ denote the optimal objective function value of LR(FLP), the Lagrangian dual problem requires maximizing $Z_{LR}(\lambda)$ over all nonnegative λ vectors. Nauss [86] employs subgradient optimization for maximizing the Lagrangian dual for the problem without single sourcing, while Bramel and Simchi-Levi [19] use the same approach for the problem with single sourcing.

For a given solution to LR(FLP), if the relaxed assignment constraints are satisfied, then we have found an optimal solution for the FLP, although this is not likely to occur in general. The resulting solution is more likely to be infeasible for the FLP, as a given customer may be assigned to multiple facilities (either violating the single-sourcing requirement, if applicable, and/or resulting in an over-assignment of a customer's demand, i.e., assigning more than 100% of a customer's demand), while another customer may not have been assigned to any facility at all. Resolving the problem of over-assigned customers is easy: we simply reduce the amount of demand assigned to facilities such that $y_j = 1$ (for example, we can retain the assignment of the customer to the open facility with the lowest value of $\hat{c}_{ij}$ in the single-sourcing case, or scale back the amount of demand assigned to multiple facilities when single sourcing is not required). Finding an assignment for an unassigned customer may not be so easy, however, as we are not guaranteed that the set of open facilities in the Lagrangian relaxation solution has sufficient capacity to accommodate all customer demands.

Bramel and Simchi-Levi [19] suggest the following heuristic approach for the problem with single-sourcing requirements. First, they solve a bin packing problem with bin capacities equal to the facility capacities and customer weights equal to their demands, d_i. The solution to the bin packing problem gives a lower bound on the number of facilities required in an optimal solution. Then, for a given λ, they solve the Lagrangian relaxation of the assignment constraints, and sort facilities in nondecreasing order of the $v(\lambda, j)$ values obtained when solving the Lagrangian subproblems. Beginning with the lowest indexed facility, they solve the 0-1 knapsack problem for this facility, with costs defined by $\bar{c}_{ij} = \hat{c}_{ij} - \lambda_i$. If the optimal knapsack solution assigns no customers, then no facility is located at the corresponding site; otherwise, the facility is opened and the associated customers from the knapsack solution are assigned. As long as unassigned customers remain, we proceed down the list of facilities, opening the next facility and solving the associated knapsack problem with all unassigned customers. If, at any point, the number of open facilities equals the minimum required number of facilities (determined by the solution of the bin packing problem) and unassigned customers remain, then these customers are assigned, one at a time, to the open facility in which they will fit at the least cost. If the set of unassigned customers cannot fit into the available capacity of the open facilities, then additional facilities are opened (with corresponding knapsack problems solved) until all customers are assigned. This solution approach works for the problem in the absence of single-sourcing requirements as well. If a customer's demand is partially assigned to a facility after solving the continuous version of the knapsack, we keep track of the remaining unassigned demand for the customer, adjust the value of $\hat{c}_{ij} = c_{ij}d_i$ to replace d_i with the number of units of unassigned demand, and proceed with the heuristic as if the remaining demand corresponds to an unassigned customer.

8.5.2 Valid inequalities for the FLP

One of the primary reasons for applying Lagrangian relaxation to the FLP is the attempt to provide a higher lower bound on the optimal solution value than that given by the LP relaxation solution. This may also be accomplished by adding valid inequalities to the formulation that eliminate segments of the LP relaxation feasible region not contained in the convex hull of mixed integer solutions. We initially consider the problem in the absence of single-sourcing requirements, noting that valid inequalities for the problem in the absence of single sourcing are also valid for the problem with single sourcing.[3] In the absence of single-sourcing requirements, the goal is to identify inequalities that are valid for the integrality requirements on the y_j variables, but that are not implied by the LP relaxation of the FLP.

We first consider the equal-capacity version of the FLP, i.e., where $k_j = k$ for all $j \in J$. For this case, letting $D(I') = \sum_{i \in I'} d_i$ for $I' \subseteq I$, the valid constraint (8.56) takes the form $k \sum_{j \in J} y_j \geq D(I)$. Using simple rounding and the integrality requirements on the y_j variables, we obtain a valid inequality

[3]Any inequality that is valid for the problem without single-sourcing requirements is clearly valid for the problem with single-sourcing requirements, as the former problem is a relaxation of the latter. However, inequalities exist that are valid for the latter problem but are not valid for the former problem. For example, the knapsack cover inequalities discussed in Section 2.6 may be applied to each facility in the single-sourcing case, although these inequalities may eliminate solutions that are feasible for the problem without single sourcing.

of the form

$$\sum_{j\in J} y_j \geq \left\lceil \frac{D(I)}{k} \right\rceil. \tag{8.66}$$

Observe that (8.66) is valid for the FLP with unequal capacities if we replace k with $k_{\max} = \max_{j\in J} k_j$. More generally, if we index facilities in nonincreasing order of capacity, where $k_{[1]} \geq k_{[2]} \geq \cdots \geq k_{[|J|]}$, and let m denote the smallest value of $\hat{\jmath}$ such that $\sum_{j=1}^{\hat{\jmath}} k_{[j]} \geq D(I)$, then we can replace the right-hand side of (8.66) with the resulting value of m. We next consider a set of inequalities that generalizes (8.66) for the equal-capacity case, and which is analogous to the valid inequality set (6.27) for the CELSP in Chapter 6.

For any subsets $I' \subseteq I$ and $J' \subseteq J$, observe that we must have $\sum_{i\in I'}\sum_{j\in J'} x_{ij} \leq k\sum_{j\in J'} y_j$ as a result of constraint set (8.33). We must also have $\sum_{i\in I'}\sum_{j\in J'} d_i x_{ij} \leq D(I')$ because $\sum_{i\in I'} d_i = D(I')$ and $x_{ij} \leq 1$ for all $i \in I$ and $j \in J$. Leung and Magnanti [74] considered these inequalities in the aggregate variable space, with $X = \sum_{i\in I'}\sum_{j\in J'} d_i x_{ij}$ and $Y = \sum_{j\in J'} y_j$; the resulting inequalities can be written as $X \leq D(I')$ and $X \leq kY$. Figure 8.2 illustrates the continuous feasible region for these constraints in the aggregate variable space. Note that the aggregate variable Y must take an integer value, and that these two constraints intersect at the point $(X, Y) = (D(I'), D(I')/k)$. Suppose that $D(I')/k$ is not an integer; then this point of intersection is not feasible for the FLP, and it is possible to add a linear constraint that eliminates this point (as well as a portion of the continuous feasible region in the neighborhood of this point) without eliminating any integer feasible points. The bold lines in Figure 8.2 indicate the feasible solutions for the FLP when Y equals the integer values $\lfloor D(I')/k \rfloor$ and $\lceil D(I')/k \rceil$.

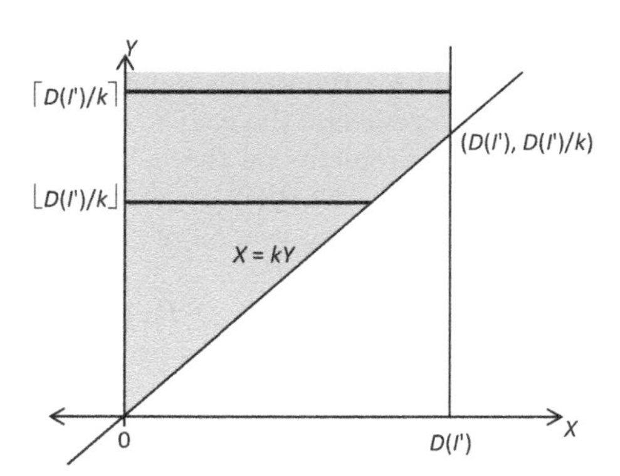

FIGURE 8.2
Illustration of the feasible region (the shaded area) for the LP relaxation of the FLP in the aggregate variable space.

Figure 8.3 illustrates how we can add a constraint that does not eliminate any integer feasible solutions. Observe that the point $(X, Y) = (D(I'), \lceil D(I')/k \rceil)$ satisfies $X \le D(I')$ at equality, while the point $(X, Y) = (k\lfloor D(I')/k \rfloor, \lfloor D(I')/k \rfloor)$ satisfies $X \le kY$ at equality. Consider the line that joins these two feasible points $(k\lfloor D(I')/k \rfloor, \lfloor D(I')/k \rfloor)$ and $(D(I'), \lceil D(I')/k \rceil)$ and note that all integer feasible points lie on one side of this line. Figure 8.3 illustrates this inequality; observe that the continuous feasible region shown in Figure 8.3 is a subset of that in Figure 8.2 as a result of the new inequality.

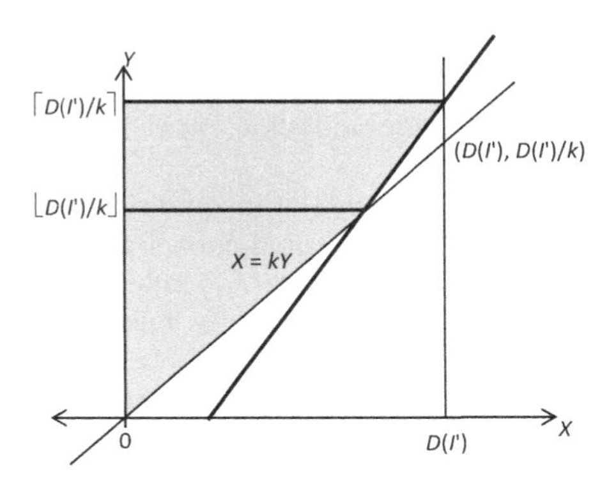

FIGURE 8.3
Illustration of valid inequality that does not eliminate any integer feasible points for the LP relaxation of the FLP in the aggregate variable space.

The slope of this line equals

$$\frac{1}{D(I') - k\lfloor D(I')/k \rfloor} = \frac{1}{r}, \tag{8.67}$$

where $r = D(I') - k\lfloor D(I')/k \rfloor = D(I')(\text{mod}k)$. This line also has a Y-intercept of $\lceil D(I')/k \rceil - D(I')/r$, implying that the line may be written in the form

$$Y = \frac{X}{r} + \left\lceil \frac{D(I')}{k} \right\rceil - \frac{D(I')}{r}, \tag{8.68}$$

or

$$X - rY = D(I') - r\left\lceil \frac{D(I')}{k} \right\rceil. \tag{8.69}$$

Because the point $(0, 0)$ must be feasible, and because $D(I') \ge r\lceil D(I')/k \rceil$,[4] the resulting inequality is written as

$$X - rY \le D(I') - r\left\lceil \frac{D(I')}{k} \right\rceil. \tag{8.70}$$

[4] $D(I') = k\lfloor D(I')/k \rfloor + r \ge r\lfloor D(I')/k \rfloor + r = r(\lfloor D(I')/k \rfloor + 1) = r\lceil D(I')/k \rceil$.

Replacing the aggregate variables with their individual terms, this becomes

$$\sum_{i\in I'}\sum_{j\in J'} x_{ij} - r\sum_{j\in J'} y_j \le D(I') - r\left\lceil \frac{D(I')}{k}\right\rceil. \tag{8.71}$$

Observe that when $I' = I$ and $J' = J$, we have $\sum_{i\in I}\sum_{j\in J} x_{ij} = D(I)$, and (8.71) reduces to (8.66). Leung and Magnanti [74] show that if $1 \le r \le k-1$ and $|J'| \ge \lceil D(I')/k\rceil$, then (8.71) serves as a facet of the convex hull of mixed integer solutions for the equal-capacity FLP (see Section 2.6 for a description of facets of a convex hull of integer solutions). They also demonstrate the equivalence of (8.71) to the single-node flow-cover facets given by Padberg et al. [89].

We next discuss an additional class of inequalities that may be very useful for the solution of the single-sourcing, equal-capacity version of the FLP for the special case in which the assignment cost ($\hat{c}_{ij}$) values are facility independent (i.e., $\hat{c}_{ij} = \hat{c}_i$ for all $j \in J$ and for each $i \in I$) and all fixed facility costs are identical (i.e., $f_j = f$ for all $j \in J$). Such problems arise in various settings in which jobs must be assigned to identical resources (e.g., machine assignment and scheduling problems). One of the challenges faced in solving the equal-capacity FLP with facility-independent assignment costs lies in the high degree of symmetry in the problem. For example, given any feasible solution to the problem when facility costs are identical, an essentially identical solution simply swaps the assignments between any two open facilities. As a result, a branch-and-bound algorithm may cycle through many permutations of customer-to-facility assignments without any real change in the resulting solution. (Note that this same issue arises in solving the generalized assignment problem with resource-independent assignment costs. Thus, the inequalities we discuss here may be useful for solving this special case of the GAP as well.) One way to mitigate this problem is via a simple set of indexing inequalities. That is, for such problems, we can, without loss of generality, assume that the number of customers assigned to facility one is at least as great as the number assigned to facility two. In general, we can assume that the number of customers assigned to facility j is at least as great as the number assigned to facility $j+1$, for all $j = 1, \ldots, |J|-1$. These indexing inequalities may be written as

$$\textstyle\sum_{i\in I} x_{ij} \ge \sum_{i\in I} x_{i,j+1} \quad j = 1, \ldots, |J|-1. \tag{8.72}$$

These inequalities eliminate the majority of cycling through identical solutions and have been found to significantly reduce the amount of time required in a branch-and-bound algorithm for practical problem classes (see, e.g., Denizel [39] and Balakrishnan and Geunes [10]).

For the general case with unequal capacities, next consider the valid inequality of the form (8.56), which is implied by the LP relaxation formulation and takes the form of a simple knapsack constraint. A subset $J' \subseteq J$ defines a cover for the demand in I if at least one of the facilities in J' must be open in order to satisfy all demand, i.e., if $\sum_{j\in J} k_j - \sum_{j\in J'} k_j < \sum_{i\in I} d_i$. A minimal cover is a cover J' such that if at least one facility $\hat{\jmath} \in J'$ is open, then this facility plus all facilities in the set $J\backslash J'$ contain sufficient capacity to meet all demand in the set I. In other words, if S is *any strict subset* of J', then $\sum_{j\in J} k_j - \sum_{j\in S} k_j \geq \sum_{i\in I} d_i$. By this definition of a minimal cover, at least one facility in a minimal cover J' must be open for any feasible solution to FLP, i.e., we must have

$$\sum_{j\in J'} y_j \geq 1 \tag{8.73}$$

if J' is a minimal cover. Aardal, Pochet, and Wolsey [2] showed that if J' is a minimal cover and $\sum_{j\in J\backslash J'} k_j + k_{\min} > \sum_{i\in I} d_i$, where $k_{\min}$ is the minimum capacity level among all facilities in J', then (8.73) serves as a facet of the convex hull of integer feasible solutions for the FLP.

Next, consider the following relaxation of the feasible region of the FLP

[CS]

$$\sum_{i\in I}\sum_{j\in J} x_{ij} = \sum_{i\in I} d_i, \tag{8.74}$$

$$\sum_{i\in I} x_{ij} \leq k_j y_j, \qquad j \in J, \tag{8.75}$$

$$x_{ij} \geq 0, \qquad i \in I, j \in J, \tag{8.76}$$

$$y_j \in \{0,1\}, \qquad j \in J, \tag{8.77}$$

which Aardal et al. [2] denote as a *cover set* (CS). A subset $J' \subset J$ is called a flow cover if $\sum_{j\in J'} k_j > \sum_{i\in I} d_i$; let $\lambda = \sum_{j\in J'} k_j - \sum_{i\in I} d_i$, where $\lambda > 0$ corresponds to the excess capacity of the facilities in the set J' with respect to the total demand $\sum_{i\in I} d_i$. The flow-cover inequality associated with the flow cover J' is written as

$$\sum_{i\in I}\sum_{j\in J'} x_{ij} + \sum_{j\in J'} (k_j - \lambda)^+ (1 - y_j) \leq \sum_{i\in I} d_i. \tag{8.78}$$

Note the similarity of (8.78) to the flow-cover inequalities (6.29) for the CELSP in Chapter 6, as the derivation of (8.78) follows similar ideas to those described in Section 6.6 for the CELSP. Aardal et al. [2] showed that if J' is a flow cover and if (i) $\max_{j\in J'} k_j > \lambda$ and (ii) $k_l < \sum_{j\in J} k_j - \sum_{i\in I} d_i$ for each $l \in J'$, then the flow-cover inequality (8.78) defines a facet of the convex hull of solutions of the FLP. They also generalize these inequalities to account for flow covers for subsets of I and J to obtain so-called effective-capacity (EC) inequalities. Aardal [1] then showed that while the separation problem[5] for these inequalities is in general $\mathcal{NP}$-Hard, when all facility capacities are

[5]The separation problem for a class of valid inequalities requires finding the most violated inequality in the class relative to a given point. That is, given a solution to a problem's relaxation and a particular type of valid inequality for the convex hull of integer solutions, we seek the inequality that maximizes the distance from the relaxation solution to the inequality.

equal, and for a fixed customer set, the separation problem for these flow-cover inequalities is polynomially solvable.

Example 8.3 Consider the problem from Example 8.1, with the addition of facility capacities $k_1 = k_4 = 20$ and $k_2 = k_3 = 15$. The LP relaxation for this instance of the FLP has an optimal solution value of 80, with $y_1 = y_2 = y_3 = y_4 = 0.5$. Consider the flow cover inequality (8.78) with the cover set $J' = J$. In this case, $\lambda = 70 - 60 = 10$, and the inequality takes the form

$$\sum_{i \in I} \sum_{j \in J} d_i x_{ij} + 10(1 - y_1) + 5(1 - y_2) + 5(1 - y_3) + 10(1 - y_4) \leq 60.$$

When this inequality is added to the LP relaxation, the optimal LP relaxation solution value increases to 100, which equals the optimal solution value of this instance of the FLP.

Alternatively, consider the generalization of the simple rounding inequalities to the case of unequal capacities. After sorting the facilities in nonincreasing order of capacity levels, the smallest value of $\hat{\jmath}$ such that $\sum_{j=1}^{\hat{\jmath}} k_{[j]} \geq D(I)$ equals 4 (where $D(I) = 60$). Thus, the inequality $\sum_{j \in J} y_j \geq 4$ is valid for this problem instance. When this inequality is added to the LP relaxation (without the addition of the previous flow cover inequality), the LP relaxation objective function value also increases to the optimal FLP objective function value of 100 for this problem instance. □

8.5.3 Approximation algorithms for the FLP

Section 8.4 discussed approximation algorithms for the MUFLP for which the resulting solution is no more than $1 + \gamma$ times the problem's optimal solution value, where γ is a fixed real number. These approximation algorithms permitted assigning each customer's demand fully to an open facility. When finite facility capacities exist, however, it is not typically possible to take an approach that assigns each customer's demand to some open facility without violating some facility's supply capacity. As a result, a different strategy is needed to arrive at an approximation algorithm for the general FLP.

Korupolu, Plaxton, and Rajaraman [70] showed that a fairly simple polynomial time local-search algorithm first proposed by Kuehn and Hamburger [71] can guarantee a solution for the metric FLP with equal capacities (without single sourcing) that is not more than $8 + \epsilon$ times the optimal solution value for a selected $\epsilon > 0$. Chudak and Williamson [28] subsequently showed that the resulting solution was guaranteed to be no more than $6(1 + \epsilon)$ times the optimal solution value, while Charikar and Guha [23] modified the algorithm and demonstrated the resulting bound to be a $3 + 2\sqrt{2} + \epsilon$ multiple. We next describe this local-search method.

For the metric FLP, if we have an optimal set of open facilities, observe that we only need to solve a transportation problem[6] in order to obtain the complete optimal solution. Our goal is therefore to find a set of open facilities that is guaranteed to at least provide a near-optimal solution. Suppose we have a set $S \subseteq J$ of open facilities (i.e., S corresponds to facilities $j \in J$ such that $y_j = 1$). The local-search method considers a set S and seeks to determine whether a slight change in the set S can provide a better solution. One such slight change would be to add some facility in $J \backslash S$ to the set S, which corresponds to opening a facility that is currently not open. We will refer to this as an *open* move. Another such change would take a single facility in S and remove it from the set, i.e., close one of the facilities in S. We will refer to this as a *close* move. The third type of change we will consider is, in effect, a combination of the previous two: we close some facility in S and open one in $J \backslash S$. We will refer to this as a *swap* move. Observe that there are $\mathcal{O}(|J|^2)$ elements of this neighborhood of solutions for a given solution defined by the set of open facilities S.

Let $c(S)$ denote the minimum cost solution when facilities in the set S are open. We only make the previously described moves if they will improve the objective function value when compared to the current solution. Suppose such a move is *allowable* only if it leads to an objective function that is reduced by at least $c(S)/p(|J|, \epsilon)$ for some polynomial function $p(|J|, \epsilon)$ of $|J|$ and ϵ. Then, assuming a sequence of allowable moves, the objective is reduced by a factor of $1 - 1/p(|J|, \epsilon)$ at each iteration, and the number of iterations required will be polynomial in $|J|$ and $1/\epsilon$. Using the supermodularity[7] of the FLP's objective function and using $p(|J|, \epsilon) \geq 8|J|/\epsilon$ with $\epsilon < 1$, Chudak and Williamson [28] demonstrate that if we have arrived at a set of open facilities S via this algorithm, and no allowable moves exist, then $c(S) \leq 6(1 + \epsilon)c(S^*)$, where S^* is the set of open facilities in an optimal solution.

Pál, Tardos, and Wexler [90] generalized this approach to allow for non-equal capacities. As before, they used three types of local moves. Their open move is identical to the previously defined open move. In addition to this, they permit two types of swaps. For a swap-close move, they close a currently open facility while (possibly) opening a subset of the previously unopened facilities to accommodate the demand from the closed facility. For a swap-open move, they open a currently unopened facility while (possibly) closing a subset of the previously open facilities. Clearly a move is only feasible if the total capacity

[6]The transportation problem is the linear programming problem obtained by fixing the binary (y_j) variables associated with opening/closing facilities in the FLP formulation, where feasibility requires that the sum of the open facility capacities is at least as great as the sum of all customer demands.

[7]Consider a function f defined over any subsets of some set S. The function f is supermodular if, for any subsets A and B of S, $f(A) + f(B) \leq f(A \cap B) + f(A \cup B)$. A function f is submodular if and only if $-f$ is supermodular. Nemhauser, Wolsey, and Fisher [88] showed that the minimum total variable supply costs ($\sum_{i \in I} \sum_{j \in J} c_{ij} x_{ij}$) are a supermodular function of the set of open facilities for the FLP. This, combined with the modularity of the total facility costs (i.e., $f(A) + f(B) = f(A \cap B) + f(A \cup B)$, where A and B are subsets of J), implies the supermodularity of the objective function of the FLP.

of open facilities is at least as large as the total demand. Because determining the best swap-close move is as hard as the FLP itself, for this type of move they provide an estimate on the cost decrease associated with the move using an upper bound on the cost of reassigning demand from the closed facility to newly opened facilities. They then show that for a candidate facility to open (close) they are able to find the minimum estimated cost among all open-swap (close-swap) moves in polynomial time in $|J|$ and the capacity of the newly opened (closed) facility. Their local-search algorithm provides a solution that is no more than $9 + \epsilon$ times the optimal solution value of the metric FLP. Zhang, Chen, and Ye [119] later introduced a different type of local move and arrived at resulting bound of $3 + 2\sqrt{2} + \epsilon$.

Levi, Shmoys, and Swamy [78] provided the first approximation algorithm for the metric FLP based on the rounded solution of the problem's LP relaxation. This algorithm uses a clustering approach that is somewhat similar to those discussed in Section 8.4 for the MUFLP. A key element of this approach lies in the use of the LP relaxation of the single-demand-node (SDN) version of the FLP, which is the special case of the FLP with only one customer with demand D. The LP relaxation of this special case of the FLP can be solved using a simple greedy approach. In particular, if we sort facilities in nondecreasing order of $f_j/k_j + c_j$ (where we have dropped the customer index i) and assign customer demand to facilities in this order until all demand is assigned (when a facility's capacity is exhausted we move to the next facility on the list), then this gives an optimal solution for the single-customer case. Observe that the resulting solution contains at most one facility for which the capacity is not fully utilized.

Another key element of the approach lies in the use of the optimal dual solution. Suppose we use the LP relaxation of the FLP(SS) formulation with the equality in constraint set (8.32) replaced by a $\geq$ inequality (note that the linear relaxation of the integrality constraints (8.34) removes the single-sourcing requirements; moreover, replacing the equality relation in (8.32) with $\geq$ does not affect the optimal solution, although the corresponding dual variables will be nonnegative). Let v_i, $i \in I$ denote the dual variables associated with constraint set (8.32).

The algorithm begins with the optimal LP relaxation solution (and the corresponding optimal dual solution) and creates clusters around cluster centers, where cluster centers correspond to customer locations. Let $\mathcal{C}$ denote the set of cluster centers (initially this set is empty; as we select customers to be cluster centers, they are added to $\mathcal{C}$). For each customer i that has not been chosen as a cluster center, Levi et al. [78] define a *closest unclustered facility set* B_i consisting of all facilities j that have not been allocated to a cluster and which have a value of c_{ij} smaller than the distance to any existing cluster center. A customer may be chosen as a cluster center only if at least half of their demand is met by facilities in B_i in the optimal LP relaxation solution. Let $\mathcal{S}$ denote the set of customers who meet this criterion. While the set $\mathcal{S}$ contains at least one element, we select from this set the customer i with the smallest value of the dual variable v_i, and center a new cluster, denoted by N_i, at this customer. All facilities in the corresponding set B_i are also allocated to this cluster. Following the creation of a cluster, we update the definitions of the sets of existing cluster centers $\mathcal{C}$, candidate cluster centers $\mathcal{S}$, and for each customer i that is not a cluster center, the set B_i. This process is repeated until the candidate cluster center set $\mathcal{S}$ is empty. If, after this step is completed, facilities remain that have not been assigned to any cluster, each remaining facility is assigned to its nearest cluster center, and the set of clusters then forms a partition of the set of facilities.

The next step in the algorithmic approach solves a problem within each cluster. Any facility such that $y_j = 1$ in the LP relaxation solution is considered open, and an SDN instance is created involving only the facilities within the cluster whose y_j variables are strictly less than one. The demand of the SDN instance is equal to the demand that is allocated to the fractionally opened facilities in the cluster in the LP relaxation solution, and this demand is located at the cluster center. Consider cluster $N_{i'}$, centered at customer i', and let $L_{i'}$ denote the set of facilities in the cluster such that $y_j < 1$. Then the demand for the SDN instance equals $\sum_{i \in I} \sum_{j \in L_{i'}} x_{ij}$, and the unit cost for serving this demand from a facility $j \in L_{i'}$ equals $c_{i'j}$. After solving the LP relaxation of this SDN instance via the greedy algorithm, any facility that satisfies a positive amount of demand in the solution is opened, and all other facilities are closed. After applying this approach to each cluster, each facility is either open or closed, and the algorithmic approach then solves a transportation problem using only the set of open facilities.

Levi et al. [78] first provide an upper bound on the solution value of the LP relaxation of the SDN instance for each cluster, and add to this an upper bound on the cost implied by fully opening the (at most one) fractionally opened facility in the LP relaxation solution of the SDN instance (assuming all facility costs are equal). They then show that the portion of the feasible solution obtained by the algorithm associated with facilities such that $y_j < 1$ in the LP relaxation of the FLP has a cost that is no more than the sum of these bounds. Next they provide a bound on the cost of the portion of the solution associated with the fully opened facilities in the LP relaxation of the FLP using LP duality. Finally, they show that the summation of these bounds is no more than 5 times the optimal solution value of the FLP assuming all equal facility opening costs, giving the first constant-factor approximation algorithm for the FLP based on linear programming.

8.6 Review

This chapter began with a profit-maximizing version of the discrete facility location problem (FLP) and used this as a starting point for showing how the FLP generalizes the previous models discussed in this book. In particular, we discussed how the knapsack problem and lot sizing problems serve as special cases of the FLP, as well as the generalized assignment problem (GAP) in the case that assignment costs are resource independent. We also highlighted the relationship between set partitioning and the FLP with single-sourcing requirements. Section 8.2.4 discussed the $\mathcal{NP}$-Hardness of the FLP, even in the case of unlimited facility capacities. Because of this, Sections 8.3 and 8.4 covered methods for obtaining heuristic solutions for the uncapacitated version of the problem via a dual-ascent algorithm and approximation algorithms with provable worst-case performance results when all distances obey the triangle inequality (the so-called metric facility location problem). Section 8.5 discussed three approaches for the problem when facilities have finite capacities. These approaches included Lagrangian relaxation (Section 8.5.1) and valid inequalities (Section 8.5.2) to obtain strong lower bounds on the optimal solution, as well as approximation algorithms for the metric FLP (Section 8.5.3).

Exercises

Ex. 8.1 — Write the dual formulation of the facility location formulation of the UELSP and describe its special structure.

Ex. 8.2 — Write out a formulation of the CELSP with single-sourcing requirements and discuss whether you think this problem might have practical applications.

Ex. 8.3 — Show that the dual ascent approach solves the dual of the facility location formulation of the UELSP and leads to a complementary primal solution satisfies the binary variable requirements.

Ex. 8.4 — Determine whether or not the problem instance in Example 8.1 is an instance of the MUFLP (metric uncapacitated facility location problem).

Ex. 8.5 — Write a set partitioning formulation of the single-sourcing version of the FLP, formulate the corresponding pricing problem, and discuss whether the pricing problem is related to any of the other problems you have studied in this book.

Ex. 8.6 — Show that the LP relaxation of the single-demand-node (SDN) version of the FLP is equivalent to a continuous knapsack problem.

9

Vehicle Routing and Traveling Salesman Problems

9.1 Introduction

The traveling salesman problem (TSP) provides a classic example of a difficult combinatorial optimization problem that can be easily explained to the layperson. A person located at a home base must visit a set of n cities and then return home, while traveling the least total distance (or incurring the lowest cost) possible. The data required in solving the problem includes the number of cities and the set of all pairwise inter-city distances. In the symmetric version of the TSP, the inter-city distance from city i to city j is the same as that from city j to city i; when considering straight-line Euclidean distances or two-way road travel distances, symmetric distances are not unlikely to apply. The general version of the TSP does not, however, impose such a symmetry requirement. Minimizing the total distance traveled serves as one possible goal; alternatively, one may wish to minimize the total cost incurred in visiting the set of cities, in which case asymmetric distances may be reasonable. For example, if air travel is involved, then it is not unlikely for the cost to travel from city i to city j to differ from the cost from j to i.

Although the idea of a person visiting a set of cities and returning home aids in conceptualizing the problem, the TSP finds application in numerous seemingly unrelated contexts. For example, consider the problem of scheduling a set of n jobs on a single machine when changeover times between jobs depend on the job sequence (for example, the time to prepare the machine for some job k after processing job i differs from the time to prepare for job j). We may think of each job as corresponding to a city and think of the changeover time when job i is performed immediately before job j as the "distance" from i to j. By adding a dummy job, which corresponds to the home city (and for which the "distance" to a job equals the setup time and the distance from a job equals any cleanup or shutdown time), the problem of sequencing the jobs with a goal of minimizing the sum of all changeover times is equivalent to the TSP. For such a problem, asymmetric "distances" are not unlikely to arise.

Another often-cited example arises in the placement of electronic components (e.g., integrated circuit chips) on a printed circuit board (PCB). Surface mount technology (SMT) component placement machines operate using a component placement head that travels above the board and drops components into place. The sequence in which components are fed to the placement head determines the sequence in which components are placed on the circuit board, which determines the total distance the head must travel above the board (and the corresponding time taken for component placement). Thus, the problem of placing the necessary components onto a circuit board corresponds to a TSP.

The TSP is something of a celebrated problem that has captured the attention and imagination of researchers around the world and is the subject of numerous books and thousands of academic articles (it is even the subject of the 2012 award-winning film *Travelling Salesman* produced by Fretboard Pictures and directed by Timothy Lanzone). The goal of this chapter is to provide an overview of this difficult problem class and to consider solution approaches that have been successfully applied to obtain high-quality solutions.

Additionally, we will discuss the related vehicle routing problem (VRP), which generalizes the TSP. In the VRP, a set of capacity-constrained vehicles must deliver goods from a distribution center (DC) to a set of geographically dispersed customers. Each customer's demands must be allocated to a set of vehicles, and the sum of the demands allocated to a vehicle may not exceed the vehicle's capacity. The objective consists of delivering all customer demands while traveling the minimum distance possible (or incurring the lowest possible travel-related cost). Given an allocation of customer demands to a vehicle, this is accomplished by solving a TSP for the vehicle, beginning at the DC and including all customer locations for customers with demands allocated to the vehicle. Thus, solving the VRP requires an ability to solve the TSP as a subproblem, and the special case of the VRP involving a single vehicle with infinite capacity is equivalent to the TSP.

9.2 The TSP Graph and Complexity

It is convenient to visualize the TSP on a complete graph consisting of $n+1$ nodes, where one of the nodes corresponds to the home location, and n additional nodes correspond to the cities to be visited. A complete graph contains an edge between each pair of nodes. Figure 9.1 shows a complete graph with nine total nodes (note that some of the edges are hidden behind others in the figure). We assume that the center node in the figure corresponds to the home node. Figure 9.1 illustrates a graph with undirected edges, which permits illustrating the symmetric TSP. The TSP with asymmetric distances requires directed edges, or arcs, with a pair of arcs (in opposite directions) between each pair of nodes. The weight associated with arc (i, j), which is directed from node i to node j, corresponds to the distance or cost associated with the arc. For the graph in Figure 9.1, beginning at the home node, observe that we have eight choices for the first city visited. Given the first city visited, we then have seven choices for the second city, and so on. As a result, a total of 8! solutions exist that visit each node exactly once and return home. In a general graph with a home city plus n additional cities, $n!$ possible solutions exist. Each such solution, which visits each node exactly once and returns home, is known as a *Hamiltonian* tour or circuit (named for the Irish mathematician Sir William Rowan Hamilton, who first posed the problem of finding a Hamiltonian tour on a graph in the 1850s). Thus, solving the TSP is equivalent to finding the least-cost Hamiltonian tour on a complete graph. Observe that if distances are symmetric, then given any tour, a tour of identical cost exists that traverses the nodes in the opposite sequence, effectively resulting in $n!/2$ unique solutions. For optimization problems in general, the number of feasible solutions does not determine the problem's difficulty (linear programs have an infinite number of solutions and can be solved in polynomial time). For $\mathcal{NP}$-Hard combinatorial optimization problems, however, finding an optimal solution may, in the worst case, require enumeration of all (or a nontrivial fraction) of the feasible solutions. The TSP is clearly in the class of $\mathcal{NP}$ problems; given a Hamiltonian tour on all of the nodes, the distance of the tour can be computed in $\mathcal{O}(n)$ time. Karp [68] showed that the general TSP is $\mathcal{NP}$-Hard, while Garey, Graham, and Johnson [52] and Papadimitriou [91] showed that this continues to hold for the symmetric TSP with Euclidean distances.

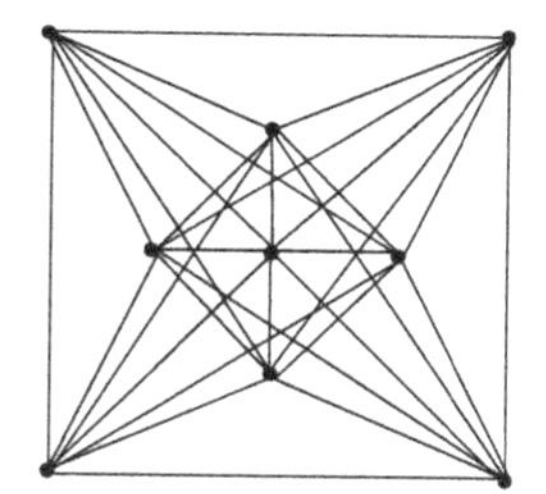

FIGURE 9.1
Complete graph with home base at center and eight additional cities.

Table 9.1 illustrates how the number of feasible solutions for the symmetric TSP grows as a function of the number of nodes n (in addition to the home city). To put this in perspective, we consider a hypothetical powerful computer.[1] At the time of this writing, the world's fastest computers can operate at a rate on the order of no more than 35 petaflops (a petaflop corresponds to 10^{15} operations per second). Let us suppose we have access to a supercomputer that can operate at 200 petaflops, which implies that we can

[1]The author would like to acknowledge Professor J. Cole Smith at the University of Florida for his input, in the form of personal communication, on providing a lower bound for the time needed to enumerate all solutions for a TSP on a hypothetical supercomputer.

Number of nodes, n	Number of tours	Supercomputer enumeration time
5	60	$<$ 1 second
10	1,814,400	$<$ 1 second
15	6.54×10^{11}	$<$ 1 second
20	1.22×10^{18}	12.165 seconds
25	7.76×10^{24}	2.46 years
30	1.33×10^{32}	$>$ 42 million years
35	5.17×10^{39}	1.64×10^{15} years

TABLE 9.1
Number of Hamiltonian tours for a symmetric TSP with n nodes plus a home base.

perform 2×10^{17} operations per second. The first time we compute the length of an initial tour, this will require $n + 1$ operations. Going from a tour to a "neighboring" tour requires four operations (we can eliminate two edges that do not touch a shared node in the initial tour, and reconnect the graph with two new edges to create a new tour; observe that there are $\mathcal{O}(n^2)$ choices of edge pairs and, for any pair of edges eliminated from a tour, there is a unique way to put the tour back together). So let us assume that computing a tour's length requires four operations for simplicity. For a symmetric TSP with $n+1$ nodes, computing all tour lengths therefore requires (more than) $4n!/2 = 2n!$ operations. The total time using our supercomputer then is $n!/10^{17}$ seconds. The third column of Table 9.1 provides a lower bound on the time it would take the supercomputer to enumerate all feasible solutions for a TSP with a home base plus n nodes. Clearly, complete enumeration is not an attractive approach for problems involving more than 20 nodes.

9.3 Formulating the TSP as an Optimization Problem

The TSP may be formulated as an optimization problem in numerous ways. If distances are asymmetric, it is convenient to define a set of directed arcs $\mathcal{A}$, with an arc existing from every node to every other node. Let $\mathcal{N}$ denote the set of nodes and let c_{ij} denote the cost or distance to travel from node i to node j for each arc $(i, j) \in \mathcal{A}$. Next, for each arc, we define a binary variable x_{ij}, which equals one if the tour visits node j immediately after node i, and equals zero otherwise. Observe that for the TSP with symmetric distances, instead of an arc set we may alternatively define an undirected edge set $\mathcal{E}$, along with a binary variable x_e for each $e \in \mathcal{E}$. Using such an approach requires explicitly keeping track of the set of edges incident to each node; that is, for each $i \in \mathcal{N}$, we define a set $\mathcal{E}_i$ containing all edges incident to node i. Because the arc-based definition can be used for problems with symmetric or asymmetric distances, we will proceed with this approach.

Given the definition of the cost c_{ij} and the binary arc selection variables x_{ij} for each $(i, j) \in \mathcal{A}$, it is easy to see that the objective of minimizing the total tour cost (or distance) can be written as

$$\text{Minimize} \sum_{(i,j)\in\mathcal{A}} c_{ij}x_{ij}, \tag{9.1}$$

an objective that is identical in form to that used for the GAP in Section 4.2 (the network induced by the GAP is, however, a bipartite network, i.e., on one side is a set of jobs, which may use the index i, and on the other side is a set of resources, which may use the index j, and arcs exist only from jobs to resources).

Given a node $i \in \mathcal{N}$, we know that for any Hamiltonian tour on the entire set of nodes, exactly one arc must be selected to enter the node and exactly one arc must be selected to leave the node. This implies the following constraints:

$$\sum_{j:(i,j)\in\mathcal{A}} x_{ij} = 1, \quad i \in \mathcal{N}, \tag{9.2}$$

$$\sum_{i:(i,j)\in\mathcal{A}} x_{ij} = 1, \quad j \in \mathcal{N}. \tag{9.3}$$

Unfortunately, constraints (9.2) and (9.3) along with the binary restrictions on the x_{ij} variables are insufficient to represent the TSP. In fact, this formulation corresponds to an instance of the standard assignment problem discussed in Section 4.2, as we can envision each city as being assigned to the next city visited in the solution. If we were to solve the problem using only the constraints we have identified thus far, and the resulting solution were to correspond to a Hamiltonian tour on all of the nodes, then the resulting solution would be optimal for the TSP (as this assignment formulation serves as a relaxation of the TSP). It is not unlikely, however, that the solution would consist of a collection of so-called *subtours*, i.e., a collection of disconnected Hamiltonian tours on subsets of the nodes, as illustrated in Figure 9.2. Providing a complete and correct formulation of the TSP therefore requires additional constraints that preclude the creation of subtours.

Numerous strategies are available for constructing such so-called *subtour elimination constraints*. Perhaps the simplest strategy to explain works as follows. Consider any set of nodes $N \subset \mathcal{N}$ of cardinality at least two (the triangular subtour in the top left corner of Figure 9.2 illustrates such a set N of cardinality three). Given such a subset N, define $A(N)$ as the set of arcs such that both the tail and head of the arc touch a node in N (i.e., $A(N)$ consists of all arcs from some node in N to another node in N). Observe that a Hamiltonian (sub)tour on the nodes in N contains exactly $|N|$ arcs. A Hamiltonian tour on all of the nodes in $\mathcal{N}$ cannot utilize $|N|$ of the arcs in $A(N)$, however, or a subtour will result. Clearly a Hamiltonian tour on all of

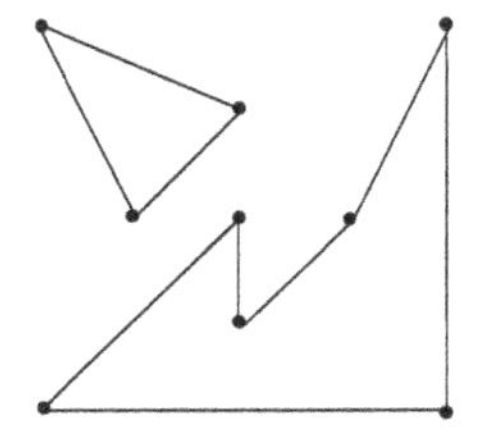

FIGURE 9.2
Two subtours on the graph from Figure 9.1 that satisfy constraints (9.2) and (9.3).

the nodes in $\mathcal{N}$ can use no more than $|N|-1$ of the arcs in $A(N)$, which leads to the validity of the following subtour elimination constraints for the TSP:

$$\sum_{(i,j)\in A(N)} x_{ij} \leq |N|-1, \quad N \subset \mathcal{N} : 2 \leq |N| \leq |\mathcal{N}|-1. \tag{9.4}$$

Grötschel and Pulleyblank [61] showed that the subtour elimination constraints (9.4) define facets of the convex hull of integer solutions for the symmetric TSP when $|\mathcal{N}| \geq 4$. Unfortunately, the number of subsets $N \subset \mathcal{N}$: $|N| \geq 3$ grows exponentially in the size of the set $\mathcal{N}$, which contains $n+1$ elements (Exercise 9.1 at the end of the chapter asks the reader to provide an expression for the number of such subsets). This implies a number of constraints that increases exponentially with n, the number of cities to be visited. Because of this, it is impractical to explicitly write out the complete set of all possible constraints of the form (9.4) for medium to large size problems, for solution via a commercial solver. Because it is easy to identify subtours in a solution to a relaxed version of the problem (e.g., the standard assignment formulation we began with before identifying the need for subtour elimination constraints), one strategy that has been successfully applied is to iteratively solve a relaxed version of the problem, identify subtours, add corresponding subtour elimination constraints, and re-solve the new relaxation. While this strategy is guaranteed to converge to an optimal solution in a finite number of steps, it may become impractical for very large problem instances (each iteration after the first requires solving a nontrivial integer program, and the number of required iterations may be large for large problems). Another way of expressing the subtour elimination constraints is by defining, for each subset $N \subset \mathcal{N}$: $|N| \geq 2$, the set of arcs such that exactly one end of the arc touches a node in N; let $\tilde{A}(N)$ denote the set of such arcs. Then, the subtour elimination constraints may be written as

$$\sum_{(i,j)\in \tilde{A}(N)} x_{ij} \geq 2, \quad N \subset \mathcal{N} : 2 \leq |N| \leq |\mathcal{N}|-1. \tag{9.5}$$

Constraint set (9.5) requires that for any set of nodes of cardinality from two to $|\mathcal{N}|-1$, at least one arc must be used to enter this set of nodes from outside the set, and at least one arc must be used to leave this set in a solution that contains no subtours.

9.4 Comb Inequalities

In order to provide strong relaxations of the TSP, researchers have identified strong inequalities and facets, in addition to the subtour elimination constraints, that aid in reducing the time required to solve the TSP via branch-and-bound. The so-called *comb* inequalities introduced by Chvátal [29] and generalized by Grötschel and Padberg [60], have been particularly successful in this regard. A comb is defined by a handle and a set of teeth. The handle, denoted by H, and each of the k teeth, denoted by T_r, $r = 1, \ldots, k$, correspond to subsets of the node set $\mathcal{N}$. Each tooth T_r has at least one node in common with the handle, and at least one node that is not part of the handle. In addition, each pair of teeth, T_r and T_s, with $r \neq s$, has no nodes in common, and we assume that at least three teeth exist in the comb, and that the number of teeth is an odd number. We focus on the symmetric version of the TSP for ease of notation, and we therefore consider the edge set $\mathcal{E}$, indexed by e, as well as the binary variable x_e, equal to one if edge e is traversed, and zero otherwise.

To derive the comb inequalities, let $E(H)$ $(E(T_r), r = 1, \ldots, k)$ denote the set of edges such that both nodes touching the edge are in H $(T_r, r = 1, \ldots, k)$. Let $\tilde{E}(H)$ $(\tilde{E}(T_r), r = 1, \ldots, k)$ denote the set of edges touching exactly one node in H $(T_r, r = 1, \ldots, k)$. Figure 9.3 illustrates the structure of the handle and teeth as well as the edge sets we have defined. Recall that for each node

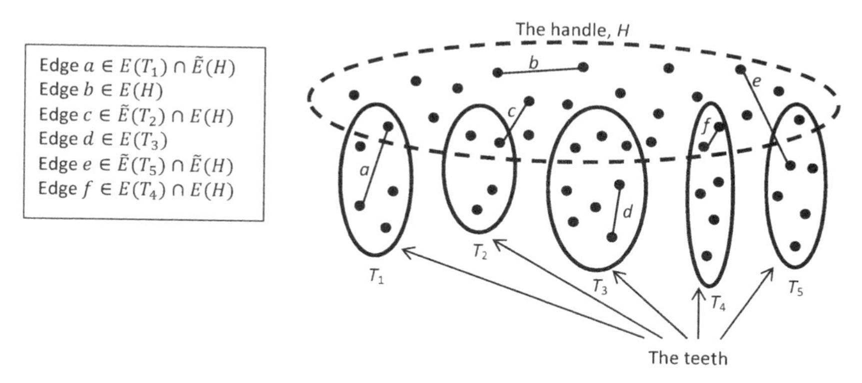

FIGURE 9.3
Illustration of set structure for comb inequalities.

$i \in \mathcal{N}$, the sum of the binary variables associated with all edges touching a node must equal two in any feasible solution, i.e., $\sum_{e \in \tilde{E}(\{i\})} x_e = 2$. Suppose that for each node in H, we multiply this equation by $1/2$, and add all such equations together. In this summation, those edges with both ends touching nodes in H, i.e., $E(H)$, will have edge variable coefficients of 1, while those with one end touching a node in H, i.e., $\tilde{E}(H)$, will have coefficients of $1/2$. The resulting equation can be written as

$$\sum_{e \in E(H)} x_e + \frac{1}{2} \sum_{e \in \tilde{E}(H)} x_e = |H|. \tag{9.6}$$

Next, note that the inequality $-1/2 x_e \leq 0$ is clearly valid for any edge $e \in \mathcal{E}$, and suppose we add this inequality to (9.6) for each edge with one end touching a node in H, but excluding any such edge that is fully contained in one of the teeth, i.e., $e \in \tilde{E}(H) \setminus \cup_{r=1}^{k} E(T_r)$, which results in

$$\sum_{e \in E(H)} x_e + \frac{1}{2} \sum_{r=1}^{k} \sum_{e \in \tilde{E}(H) \cap E(T_r)} x_e \leq |H|. \tag{9.7}$$

Next, consider the subtour elimination constraints of the form (9.4) applied to each of the subsets T_r, $H \cap T_r$, and $T_r \setminus H$ for $r = 1, \ldots, k$ (observe that $T_r = \{H \cap T_r\} \cup \{T_r \setminus H\}$, and the dashed oval in Figure 9.3 partitions T_r into the two disjoint subsets $\{H \cap T_r\}$ and $\{T_r \setminus H\}$). Suppose we take each of these subtour elimination constraints, multiply them by $1/2$, and add them together. After this summation, those edges with both ends touching nodes in T_r, $r = 1, \ldots, k$, which do not "cross the boundary of the handle" (the dashed oval in Figure 9.3) will have edge variable coefficients of one; this is the set of edges $E(T_r) \setminus \tilde{E}(H)$. The remaining set of edges in $E(T_r)$, which is the set $E(T_r) \cap \tilde{E}(H)$, will have edge variable coefficients of $1/2$. The resulting inequality is written as

$$\frac{1}{2} \sum_{r=1}^{k} \sum_{e \in \tilde{E}(H) \cap E(T_r)} x_e + \sum_{r=1}^{k} \sum_{e \in E(T_r) \setminus \tilde{E}(H)} x_e \leq \sum_{r=1}^{k} |T_r| - \frac{3}{2} \sum_{r=1}^{k} 1. \tag{9.8}$$

Aggregating (9.7) and (9.8), and noting that $\{\tilde{E}(H) \cap E(T_r)\} \cup \{E(T_r) \setminus \tilde{E}(H)\} = E(T_r)$, gives

$$\sum_{e \in E(H)} x_e + \sum_{r=1}^{k} \sum_{e \in E(T_r)} x_e \leq |H| + \sum_{r=1}^{k} (|T_r| - 1) - \frac{k}{2}. \tag{9.9}$$

Recognizing that k is odd (and $k/2$ is therefore fractional), we can reduce the right-hand side by $1/2$ giving the comb inequality ([60]):

$$\sum_{e \in E(H)} x_e + \sum_{r=1}^{k} \sum_{e \in E(T_r)} x_e \leq |H| + \sum_{r=1}^{k} (|T_r| - 1) - \frac{k+1}{2}. \tag{9.10}$$

Exercise 9.2 asks the reader to verify the validity of the comb inequalities (9.10). Grötschel and Pulleyblank [61] showed that the comb inequality (9.10) is a facet of the convex hull of *path systems* for the TSP when $\mathcal{N} \geq 6$ (a path system is a set of paths on the nodes in the TSP such that no two paths share a node; such a set of paths can be connected by edges to form a tour). They also further generalized the comb inequalities to the case of *clique trees* consisting of connected sets of handles and teeth such that no two teeth intersect and no two handles intersect, but a tooth may connect to multiple handles. Each tooth has at least two nodes and at most $|\mathcal{N}| - 2$ nodes, along with at least one node that is not part of any handle. Every handle has an odd number of teeth greater than or equal to three, and if a handle and tooth have an intersection, the removal of the nodes in this intersection results in more connected components than are contained in the original clique tree. Grötschel and Pulleyblank [61] show that the resulting clique tree inequalities serve as facets for the symmetric TSP. Goemans [57] discusses additional classes of inequalities and facets for the TSP and provides key results on worst-case bounds on the performance of these classes of inequalities in strengthening relaxations for the TSP.

9.5 Heuristic Solutions for the TSP

Because the TSP is so widely applicable to practical contexts that may require fast solutions, researchers have developed a variety of heuristic solution strategies for obtaining good solutions quickly. We assume that the heuristics described in this section are applied to a metric TSP, wherein all distances satisfy the triangle inequality (i.e., $c_{ik} \leq c_{ij} + c_{jk}$ for all nodes, i, j, k).

9.5.1 Nearest neighbor heuristic

Perhaps the simplest heuristic approach, known as the *nearest neighbor* method, is effectively a greedy approach, which corresponds to what a decision maker might do based on intuition. This method simply starts at the home base and, at each step, moves to the nearest location that has not yet been visited, breaking ties arbitrarily.

Example 9.1 Figure 9.4 shows a nearest neighbor solution for the graph shown in Figure 9.1, assuming the center node corresponds to the home base. We assume that each pair of four nodes equidistant from the center node forms a square centered at the home base, where the inner square has side length L_1 and the outer square has side length L_2, with $L_2 > L' > L_1$. As we will discuss shortly, whether or not the resulting solution is optimal depends on the relative values of L_1 and L_2. □

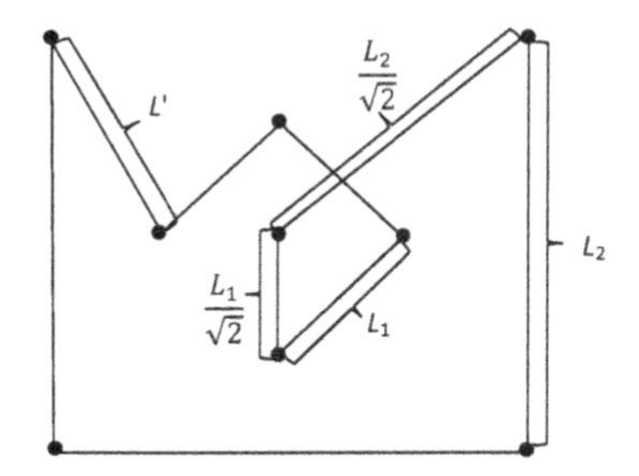

FIGURE 9.4
Nearest neighbor solution for graph in Figure 9.1.

Unfortunately the nearest neighbor method may perform poorly for certain graph types. We next illustrate this via an example.

Example 9.2 Consider the problem instance depicted in Figure 9.5, where node 0 corresponds to the home location and the distances are shown in the figure.

FIGURE 9.5
Set of nodes corresponding to a TSP instance on a line.

Figure 9.6 shows a nearest neighbor solution to the problem depicted in Figure 9.5. The distance traveled in this solution equals 42, while the minimum cost solution, which is obtained by traveling from node 0 to each of the extreme nodes (the left-most and right-most nodes in the figure) and back to node 0, equals 32.

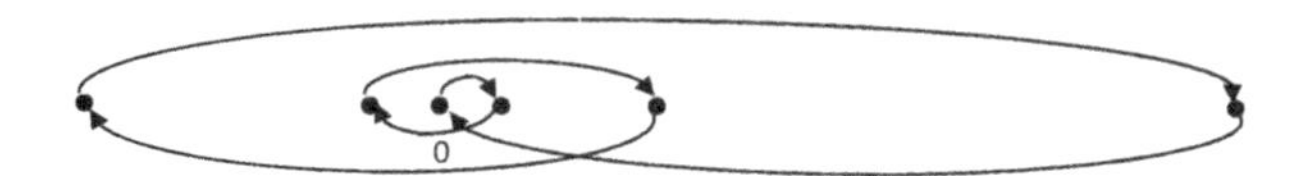

FIGURE 9.6
Nearest neighbor solution for the TSP instance on a line shown in Figure 9.5.

If we generalize the construction of the graph in Figure 9.5 to contain n nodes (repeating the pattern shown in the figure, where the greatest arc length equals 2^{n-3}), then it is possible to show that the nearest neighbor solution for the alternating pattern shown in Figure 9.6 has length equal to $\sum_{i=1}^{\lfloor n/2 \rfloor} 2^{n-2i+1}$, while the optimal solution has length 2^{n-1}. This implies a relative performance (the ratio of the nearest neighbor solution to the optimal solution) equal to

$$\sum_{i=1}^{\lfloor n/2 \rfloor} 2^{2-2i} = 4 \sum_{i=1}^{\lfloor n/2 \rfloor} \left(\frac{1}{4}\right)^i. \tag{9.11}$$

If we take the limit of the right-hand side of (9.11), which is a convergent geometric series, as $n \to \infty$, we obtain a limit of 16/3, indicating that the particular graph and pattern we have illustrated does lead to a worst-case performance bound for the nearest neighbor approach as the number of nodes, n, increases. □

Rosenkrantz, Stearns, and Lewis [96] showed that instances of the TSP exist in which the ratio of the nearest neighbor solution to the optimal solution is $\mathcal{O}(\log n)$, implying that no worst-case performance bound exists in general for the nearest neighbor method that is independent of the number of nodes (and, therefore, independent of the problem size).

To improve the nearest neighbor method, Rosenkrantz et al. [96] suggest a nearest insertion heuristic, which builds upon an existing Hamiltonian tour on a subset of the nodes (a subtour), and which we next describe. Suppose we have a Hamiltonian tour on a subset of nodes $N' \subset \mathcal{N}$. Determine the node k in $\mathcal{N} \backslash N'$ with the shortest distance to a node in N'. For each edge (i, j) in the subtour, compute the increase in total distance if node k is inserted between nodes i and j (and the subtour size is therefore increased by one node), i.e., $c_{ik} + c_{kj} - c_{ij}$. For the edge (i, j) that gives the minimum such insertion value, insert k into the tour between i and j, and let $N' = N' \cup \{k\}$. Repeat this process until $N' = \mathcal{N}$. Bramel and Simchi-Levi [19] showed that this heuristic gives a solution of no more than twice the optimal solution value.

9.5.2 The sweep method

Another simple construction method is known as the *sweep* method, provided by Gillett and Miller [56] for the VRP. This method starts with a central node and extends a line (or sweep arm) from the node, which is at least as long as the straight-line distance from the central node to the node that is furthest from this node. The line then sweeps through a complete circle in either direction and adds nodes to the tour in the sequence in which the sweep arm touches the node.

Example 9.3 Figure 9.7 shows the results of application of the sweep method to the graph in Figure 9.1. A sweep arm of length $L_2/\sqrt{2}$, which is dropped vertically from the center node and sweeps clockwise, leads to the solution shown in the figure. □

Clearly the choice of the center location, the initial sweep arm position, and the sweep direction all influence the quality of the solution obtained. Because this method can be applied quickly, it may be restarted multiple times with

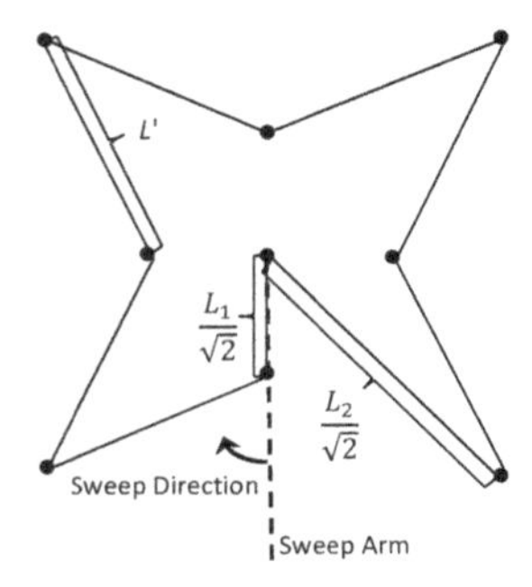

FIGURE 9.7
Sweep method solution for the TSP instance shown in Figure 9.1.

varying initial sweep arm positions and sweep directions in order to obtain multiple solutions, from which the best may be selected.

Example 9.4 Figure 9.8 shows the results of an application of the sweep method to the line graph from Figure 9.5, using a sweep arm extending from node 0 (for this graph, the initial sweep arm position and sweep direction do not affect the solution). It turns out that for this graph, the sweep method provides an optimal solution. □

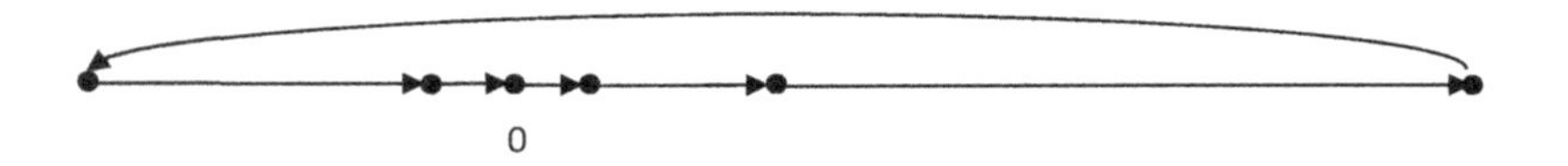

FIGURE 9.8
Sweep method solution for the TSP instance on a line shown in Figure 9.5.

The sweep method, like the nearest neighbor method, can perform poorly for certain graph types. For example, it is not clear whether the solution shown in Figure 9.7 is a high quality solution. This depends on the relative values of L_1 and L_2. Assuming $L_1 = 1$, then the value of the nearest neighbor solution, illustrated in Figure 9.4, is less than the value of the sweep method solution, shown in Figure 9.7, whenever $L_2 < 2L' - 1$, which turns out to correspond to the condition $L_2 > 1 + \sqrt{2} + \sqrt{2(1+\sqrt{2})}$, and the cost difference $(2L' - L_2 - 1)$ is increasing and convex for all $L_2 > 1 + \sqrt{2} + \sqrt{2(1+\sqrt{2})}$. The sweep method results in multiple traversals from nodes in the "inner square" to nodes in the "outer square" and back again. To overcome this potential pathological behavior of the sweep method, we may consider using the sweep method within concentric disks, i.e., nodes are visited in the sequence in which the sweep arm hits them within each of a set of concentric disks (see, for example, Haimovich and Rinnooy Kan [64]).

Example 9.5 Figure 9.9 illustrates the application of the sweep method within concentric disks to the graph in Figure 9.1. The inner disk is centered at the home node with a radius between $L_1/\sqrt{2}$ and $L_2/\sqrt{2}$, while the outer disk has radius at least $L_2/\sqrt{2}$. The nodes within each disk are visited in the order in which the sweep arm hits them, and the disks are connected by a single arc from the inner disk to the outer disk, while the last node visited in the outer disk is connected to the home base. Observe that the solution obtained via this method is, for the graph illustrated, identical to the nearest neighbor solution (although in general this is not likely to be the case). □

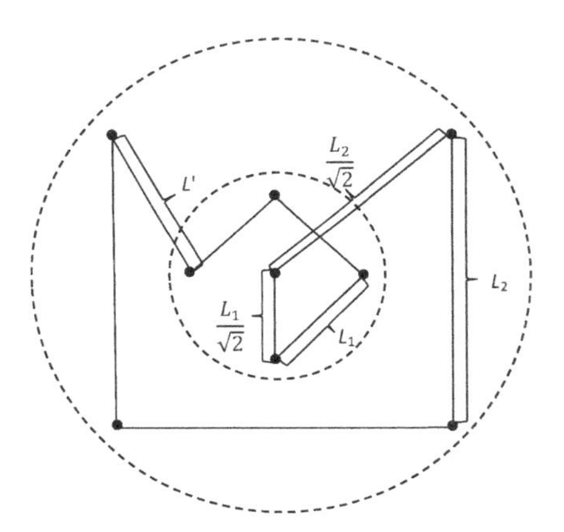

FIGURE 9.9
Sweep method solution within concentric disks for the TSP instance shown in Figure 9.1.

Clearly this approach can be applied to multiple concentric disks and, like the sweep method, multiple different sweep arm angles may be tested within each disk.

9.5.3 Minimum spanning tree based methods

Another class of heuristics with constant-factor approximation bounds is based on a minimum spanning tree solution. A spanning tree on a connected graph with node set $\mathcal{N}$ and edge set $\mathcal{E}$ is a subgraph containing all nodes in $\mathcal{N}$ that is connected and contains no cycles. The minimum spanning tree (MST) problem requires determining the spanning tree with the minimum value of the sum of all edge weights. This problem can be solved by a simple greedy algorithm (see, e.g., [6]), which works as follows. First create a sorted list of all edges in nondecreasing order of edge weights, breaking ties arbitrarily. Process edges in this order, adding an edge to the solution as long as it does not create a cycle (if an edge forms a cycle, skip this edge and move down the list in the sorted order).

Note that the MST solution cannot be larger than the optimal TSP solution, as the removal of one edge from the TSP solution produces a spanning tree. Suppose we take the MST and duplicate each edge. The resulting graph allows for a *Eulerian* tour, i.e., a solution beginning at the home node that traverses each edge exactly once and returns to the home node. The length of this Eulerian tour is twice that of the MST solution value, which is bounded by twice the optimal TSP solution. Although this solution does not correspond to a Hamiltonian tour, a Hamiltonian tour may be created by taking shortcuts. For example, suppose our Eulerian tour goes from node i to node j, back to node i, and then to node k. The cost of this segment of the Eulerian tour equals $2c_{ij} + c_{ik}$. If we instead take a shortcut by going directly to node k after node j, then the cost is reduced to $c_{ij} + c_{jk}$ (this argument can be extended to the case in which node i is followed by a sequence of nodes before reaching node j, then returns to node i before visiting node k). Because of the triangle inequality, $c_{jk} < c_{ij} + c_{ik}$, and the value of the resulting Hamiltonian tour created by shortcuts is guaranteed to be less than twice the MST solution value. This heuristic approach therefore has a worst-case performance bound of twice the optimal solution value of the TSP.

Example 9.6 Consider the graph in Figure 9.1; the MST for this graph is shown on the left in Figure 9.10. If we duplicate the edges in the MST, we arrive at the graph shown on the right in Figure 9.10, which also shows the sequence of edge visits in a Eulerian tour. We next shortcut the Eulerian tour to eliminate multiple visits to a node. The graph on the left in Figure 9.11 shows the shortcuts taken (illustrated by the dashed edges), resulting in the solution shown in the graph on the right in Figure 9.11. □

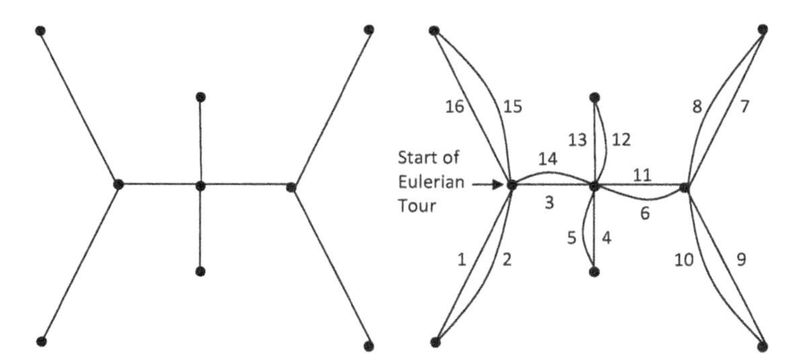

FIGURE 9.10
Minimum spanning tree for graph from Figure 9.1.

Christofides [25] improved upon this approach as follows. First note that after the creation of the MST, the number of nodes of odd degree is even. Instead of duplicating the edges in the MST, Christofides [25] solves a minimum weight matching on the nodes of odd degree in the MST, and adds the

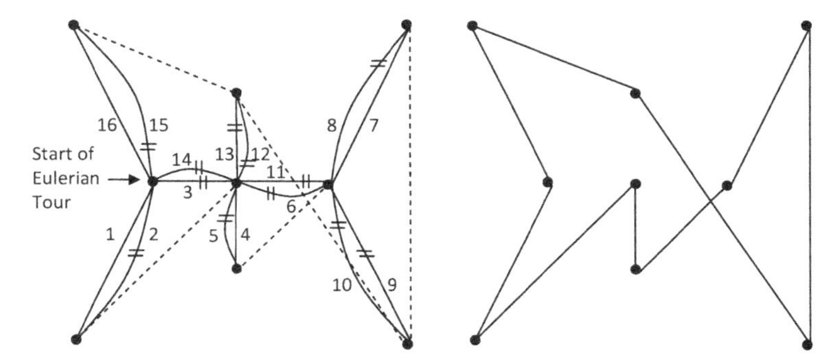

FIGURE 9.11
TSP solution from duplicating edges in MST, walking a Eulerian tour in the graph, and taking shortcuts.

edges in the resulting solution to the MST to form a graph that contains a Eulerian tour. The sum of the edge weights in this Eulerian tour equals the cost of the MST plus the minimum weight matching cost, which is bounded by the minimum weight matching cost plus the optimal TSP solution value. Thus, if we eliminate multiple visits to a node by taking shortcuts as described previously (in order to create a Hamiltonian tour), the resulting Hamiltonian tour has a corresponding solution value that is no more than the optimal TSP cost plus the cost of the minimum weight matching on the nodes of odd degree. We can bound the cost of this minimum weight matching as follows. Consider the optimal Hamiltonian tour on the set of nodes of odd degree in the optimal MST solution, which we denote as O. Note that by the triangle inequality, the cost of this tour is bounded by the optimal TSP solution in the network. Suppose without loss of generality that the optimal Hamiltonian tour on the set of nodes in O visits edges in the sequence $e_1, e_2, \ldots, e_k$, where $k = |O|$. Note that the set of edges $\{e_1, e_3, \ldots, e_{k-3}, e_{k-1}\}$ corresponds to a feasible matching with respect to the nodes in O, as does the set of edges $\{e_2, e_4, \ldots, e_{k-2}, e_k\}$. The sum of the weights of both of these matching solutions is no more than the optimal TSP solution value. The minimum value between these two matching solutions, therefore, can be no more than 1/2 the optimal TSP solution value. Because the minimum weight matching solution on the nodes in O is no larger than the minimum between these two matching solutions, this implies that this minimum weight matching solution on the nodes in O is no larger than 1/2 the optimal TSP solution value. This further implies that the solution provided by this heuristic approach is bounded by 3/2 times the optimal TSP solution value.

9.5.4 Local improvement methods

Some of the most successfully applied solution methods for the TSP rely on beginning with a feasible solution and then making relatively simple alterations to the solution in order to obtain a new solution (see, e.g., [41]). By defining a particular class of alterations to a solution, we define a local neighborhood of solutions that may result from the particular class of alterations. Given a solution, the goal is then typically to find the element within the local neighborhood with the highest degree of improvement over the existing solution, i.e., the minimum cost solution in the local neighborhood. Upon obtaining the new solution, the process is repeated with the newly defined local neighborhood, until a locally optimal solution is obtained, i.e., a solution in which no better solution exists within the local neighborhood.

Croes [33] appears to be the first work to apply this sort of approach, having defined what is now known as the 2-opt exchange method for the symmetric TSP. The 2-opt exchange works by eliminating two arcs from a solution and then piecing the solution back together in a different (unique) way. For example, suppose a solution uses arcs (i_1, j_1) and (i_2, j_2) where i_1, i_2, j_1, and j_2 correspond to four distinct nodes. If we eliminate these arcs, we can piece the solution back together using arcs (i_1, j_2) and (i_2, j_1). If $c_{i_1 j_2} + c_{i_2 j_1} < c_{i_1 j_1} + c_{i_2 j_2}$, then the new solution improves over the initial solution. Since there are $\mathcal{O}(n^2)$ choices of arc pairs, each iteration can be evaluated in polynomial time. Because of the algorithm's speed, the procedure may be restarted multiple times with different initial solutions, leading to different locally optimal solutions. Lin and Kernighan [80] generalized this approach to allow for k-opt exchanges, where k denotes the number of arcs removed from the initial solution (and replaced with k other arcs). For $k > 2$, however, there is no longer a unique way to piece the solution back together, and multiple options must be considered (and the number of such options increases exponentially in k). In the Lin and Kernighan approach, multiple values of k are evaluated at each iteration, before making a change to the original solution. While the Lin and Kernighan method has demonstrated extremely effective performance in practice, Papadimitriou, Schäffer, and Yannakakis [92] showed that it falls into a class of local search problems known as Polynomial Local Search-Complete, or PLS-Complete, which is the most difficult class of local optimization problems.

9.6 The Vehicle Routing Problem

The vehicle routing problem (VRP), as noted earlier, considers a set $\mathcal{V}$ of vehicles located at a home base, which we will call the distribution center (DC). Associated with each $v \in \mathcal{V}$ is a delivery capacity K. As in the TSP, we consider a node set $\mathcal{N}$, as well as an inter-node distance (or cost) c_{ij} corresponding to the distance (or cost) between nodes i and j for each node pair $i, j \in \mathcal{N}$. We assume without loss of generality that node 0 corresponds to the DC and that n customer nodes exist. Associated with each non-zero (customer) node is a demand quantity; let d_i denote the demand associated with node $i \in \mathcal{N}\backslash\{0\}$. The version of the VRP we will consider assumes that a customer's demand may not be split among multiple delivery vehicles. This is analogous to the single-sourcing version of the facility location problem in Chapter 8. While permitting delivery splitting among multiple vehicles leads to a larger feasible region (and, therefore, corresponds to a relaxation of the problem without delivery splitting), in practice, customers often prefer a single consolidated delivery in order to simplify the management of receiving operations.

The goal of the VRP is to deliver all customer demands while incurring the minimum possible delivery cost or traveling the minimum total distance. The VRP essentially combines two types of decisions. The first type of decision requires determining which customers will be assigned to a common vehicle. Given a set of customers assigned to a vehicle, we must then determine the sequence in which the vehicle delivers to customers. This latter decision corresponds to solving a TSP involving the DC and the subset of customers assigned to the vehicle. The VRP therefore generalizes the TSP, implying that the VRP is, in general, at least as hard a problem as the TSP, and often necessitates the use of heuristic solution methods in practice. Note that the approach of assigning customers to vehicles and then solving a TSP for each vehicle is known as a "cluster-first, route-second" heuristic solution approach. An alternative family of heuristics operates using a "route-first, cluster-second" approach (see, e.g., Beasley [16]), wherein a TSP may be solved on all nodes, and the resulting solution is then broken up into individual vehicle routes. We will subsequently describe two often-used cluster-first, route-second heuristic approaches, as well as an exact optimization method based on column generation for a set partitioning formulation.

9.6.1 Exact solution of the VRP via branch-and-price

Because of the assignment structure of the VRP without delivery splitting, we can view the problem as a set partitioning problem. That is, we must partition the set of customers such that a single vehicle serves each customer subset. We will therefore assume that the size of the set of vehicles $\mathcal{V}$ is sufficient to ensure that a feasible solution exists (it is sufficient, and convenient, to assume that the vehicle capacity is at least as great as the largest customer demand, and that the number of vehicles equals the number of customers). Suppose we can enumerate all feasible customer subsets $\mathcal{F}$, i.e., for any set of customers $S \in \mathcal{F}$

$$\sum_{i \in S} d_i \leq K, \quad S \in \mathcal{F}. \tag{9.12}$$

Let C_S denote the minimum distance traveled (or cost incurred) in delivering to the subset of customers S, and let a_S denote an n-dimensional column vector such that the ith element of a_S equals one if customer i is included in S, and equals zero otherwise. Letting y_S denote a binary variable equal to one if subset S is selected (and zero otherwise) for each $S \in \mathcal{F}$, then we can formulate the VRP as an equivalent set partitioning problem using the SP formulation (3.1)–(3.3). Unfortunately, the number of elements of $\mathcal{F}$ may be extremely large, prohibiting explicitly expressing such an SP formulation in its entirety. In order to attempt to overcome this limitation, we consider the column-generation and branch-and-price methods described in Sections 3.3.2 and 3.3.3, respectively.

Recall that the branch-and-price method begins by solving the LP relaxation of the set partitioning problem at the root node via column generation. Starting with a subset of the columns that guarantees a feasible solution, and an LP relaxation to this restricted problem, the pricing problem is solved in order to identify whether a column exists that prices out, i.e., a column that has an attractive reduced cost. If this is the case, the column is added to the restricted problem and the LP relaxation is re-solved. When no columns price out, we have an optimal solution to the LP relaxation of the set partitioning problem. If this solution is binary, then it solves the set partitioning problem. Otherwise, some fractional variables exist, and we branch on a fractional variable (or a subset of fractional variables, as discussed in Section 3.3.3). The general form of the pricing problem takes the form of problem PP in Equations (3.4)–(3.6), which is reproduced below for convenience.

$$\textbf{[PP]} \qquad \text{Minimize} \qquad C_S(a_S) - \pi a_S \tag{9.13}$$

$$\text{Subject to:} \qquad a_{S,i} \in \{0,1\}, \qquad i = 1, \ldots, n, \tag{9.14}$$

$$S \in \mathcal{F}. \tag{9.15}$$

For the VRP, $C_S(a_S)$ corresponds to a minimum distance (or cost) TSP solution that includes the customers in the subset S as well as the DC (node 0), and π (the set of dual variables associated with the assignment constraints in the set partitioning formulation) is equivalent to a vector of rewards for including customers in the subset. The resulting pricing problem corresponds to a prize-collecting TSP (see Balas [11]) with an additional constraint requiring that the sum of the customer demands on the route cannot exceed the vehicle capacity K. That is, we seek a Hamiltonian tour on a subset of customers whose demands do not exceed K, which maximizes the sum of the rewards minus the total distance traveled (or cost incurred). Foster and Ryan [51] solve this pricing problem via dynamic programming, despite the fact that the pricing problem is, unfortunately, an $\mathcal{NP}$-Hard optimization problem. Desrochers, Desrosiers, and Solomon [40] suggest using an alternative formulation of the set partitioning problem, which leads to a pricing problem that may be solved in pseudopolynomial time. This alternative set partitioning (ASP) problem is formulated as

[ASP] Minimize $\sum_{S \in \mathcal{F}} C_S(a_S)y_S$ (9.16)

Subject to: $\sum_{S \in \mathcal{F}} a_S y_S \geq e^{|I|}$, (9.17)

$y_S \in \{0,1\}, \qquad S \in \mathcal{F},$ (9.18)

where a_S, instead of being defined as a binary vector, is now defined as a vector of nonnegative integers such that $a_{S,i}$ equals the number of times customer i is visited in the route corresponding to subset S. This ASP formulation, therefore, permits routes in which a customer is visited multiple times within the route. While an optimal solution will never select such routes, this leads to pricing problems that are easier to solve as shortest path problems via dynamic programming. This shortest path problem computes the shortest path starting at the depot and ending at node i with a vehicle load equal to k ($\leq K$); for each (i,k) pair, this requires $\mathcal{O}(n^2K)$ computations (see Bramel and Simchi-Levi [19]). Desrochers et al. [40] provide results of a broad set of computational experiments illustrating the effectiveness of this approach for solving the VRP with additional time window constraints (in which each customer may only be visiting during certain customer-specific time intervals).

9.6.2 A GAP-based heuristic solution approach for the VRP

Fisher and Jaikumar [47] provided an effective heuristic solution approach by casting the VRP as a generalized assignment problem (GAP). In this version of the GAP, the set of vehicles $\mathcal{V}$ serves as the set of resources to which customer demands (the items) must be assigned. Because the GAP permits non-identical resource capacities, we let K_v denote the capacity of vehicle $v \in \mathcal{V}$. As in the definition of the GAP in Section 4.2, we define a binary variable x_{iv} for each $i \in \mathcal{N}$ and $v \in \mathcal{V}$, where $x_{iv} = 1$ if customer i is assigned to vehicle v, and $x_{iv} = 0$ otherwise (in this GAP formulation, the set $\mathcal{N}$ replaces the set I while the set $\mathcal{V}$ replaces the set J in the formulation in Section 4.2). Instead of defining b_{ij} as the amount of resource j consumed by item i, we let d_i denote the amount of vehicle capacity consumed by customer i, independent of the vehicle to which the customer is assigned. Moreover, we cannot use a simple assignment cost of c_{ij} corresponding to the cost of assigning item i to resource j. Instead, the cost of assignments to a vehicle is a nonlinear function $f(x_{\cdot v})$, where $x_{\cdot v}$ denotes an $|\mathcal{N}|$-vector of x_{iv} values. The function $f(x_{\cdot v})$ corresponds to the least cost Hamiltonian tour containing all customers i such that $x_{iv} = 1$, as well as the DC. Clearly, the computation of $f(x_{\cdot v})$ requires solving a TSP and is, therefore, an $\mathcal{NP}$-Hard optimization problem. Because of this, Fisher and Jaikumar [47] use a linear approximation for the cost of assigning customer i to vehicle v, which we denote as $\tilde{c}_{iv}$.

These approximate assignment costs are obtained after first identifying, for each vehicle (which corresponds to a route), a route *seed* point. This seed point location may correspond to a customer location, or may be any point on the plane selected according to a particular rule or policy (clearly the quality of solution and approximation depends on the choice of seed points). Fisher and Jaikumar [47] discuss a way to select seed points via the creation of a set of cones originating at the DC, one for each customer; contiguous *customer cones* are then aggregated into route, or *vehicle cones*, and the vth seed point is located on the ray that bisects the vth vehicle cone. We illustrate this creation of cones via an example that uses the graph shown in Figure 9.1. Suppose that the center node corresponds to the DC and we have four vehicles, each with a capacity of 125 units. The customer demands are shown next to each node in Figure 9.12 (note that the total demand equals 400 units). We initially create a cone for each customer, where the half-ray determining the boundary between any two customers occurs halfway between the customers, as shown in Figure 9.12. We envision each customer's demand as being uniformly distributed throughout the cone, and suppose we sweep clockwise through the customer cones, in order to create four vehicle cones, one for each vehicle. To create these vehicle cones, note that the average demand carried per vehicle will equal 100 units. Therefore, we rotate the sweep arm, and create a vehicle cone boundary each time we have swept through 100 units of demand (for example, 1/3 of a sweep through a customer cone accumulates 1/3 of the customer's demand). This results in the four vehicle cones shown in Figure 9.13 indicated by the dashed half-rays. For each of the vehicle cones, we create a seed point on the half-ray that bisects the vehicle cone, at a distance equal to the straight-line distance of the furthest customer in the cone, as shown in Figure 9.13.

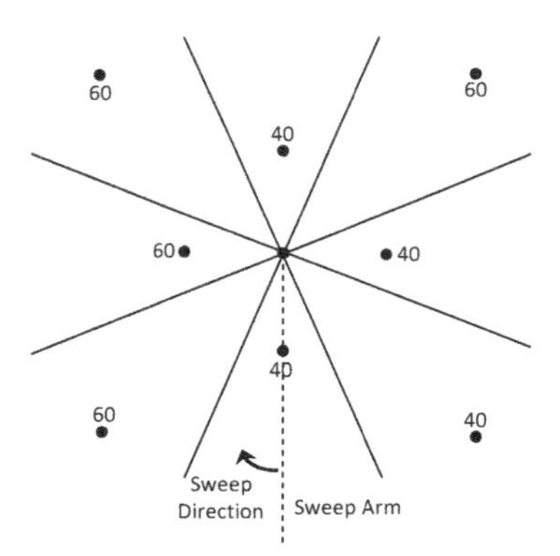

FIGURE 9.12
Illustration of customer cones for the graph from Figure 9.1.

Suppose we have identified such a seed point i_v for each $v \in \mathcal{V}$ (where node i_v is located at the seed point for route v). We assume for ease of explanation that distances are symmetric, although the analysis extends easily to asymmetric distances. To quantify the approximate assignment cost for customer i and route v, Fisher and Jaikumar [47] use the cost of inserting customer i

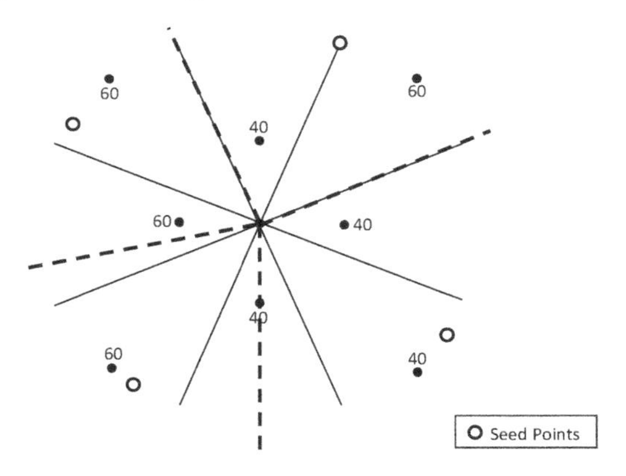

FIGURE 9.13
Illustration of vehicle cones obtained for the example shown in Figure 9.12.

into a route that contains only the DC and the seed point i_v, i.e.,

$$\tilde{c}_{iv} = c_{0i} + c_{i,i_v} - c_{0i_v}. \tag{9.19}$$

The resulting formulation of the GAP, denoted as the VRPGAP, is written as

$$\textbf{[VRPGAP]} \quad \text{Minimize} \quad \sum_{i \in I} \sum_{j \in J} \tilde{c}_{iv} x_{iv} \tag{9.20}$$

$$\text{Subject to: } \textstyle\sum_{i \in I} d_i x_{iv} \leq K_v, \quad v \in \mathcal{V}, \tag{9.21}$$

$$\textstyle\sum_{v \in \mathcal{V}} x_{iv} = 1, \quad i \in \mathcal{N} \setminus \{0\}, \tag{9.22}$$

$$x_{iv} \in \{0, 1\}, \quad i \in \mathcal{N} \setminus \{0\}, v \in \mathcal{V}. \tag{9.23}$$

Observe that the VRPGAP is equivalent to a discrete facility location problem with single-sourcing requirements and zero fixed facility costs (FLP(SS); see Section 8.2.3). After solving the VRPGAP (perhaps via one of the methods discussed in Chapter 4), we have an assignment of customers to vehicles. For each vehicle and the customers assigned to the vehicle, we then require the application of a method for solving the TSP containing the DC and the customers assigned to the vehicle (several such methods have been described earlier in this chapter). As a result, this approach falls into the category of cluster-first, route-second heuristic methods.

9.6.3 The Clarke-Wright savings heuristic method

Perhaps the most well-known heuristic solution method for the VRP is also a cluster-first, route-second method developed by Clarke and Wright [30], known appropriately as the Clarke-Wright savings method. This simple method for clustering customers is the original version of the type of calculation Fisher and Jaikumar [47] used in Equation (9.19) to approximate the cost of assigning a customer to a seed point. The method essentially begins with a very poor solution in which each customer is assigned to a single vehicle (this initial solution is not only likely to be a poor solution, but is also infeasible when the number of vehicles is less than the number of customers). For each pair of customers, we compute the savings in cost (or distance) that results if the two customers are assigned to the same route. For customers i and j, this savings equals (again assuming symmetric distances for simplicity)

$$s_{ij} = c_{0i} + c_{0j} - c_{ij}. \tag{9.24}$$

After calculating the savings for each customer pair, we create an ordered list of these savings values from largest to smallest. We then process customer pairs in this order. For example, beginning with the first pair (i_1, j_1), and assuming each vehicle has capacity K, if $d_{i_1} + d_{j_1} \leq K$, then customers i_1 and j_1 will be assigned to the same vehicle. If this is the case, then customers i_1 and j_1 are effectively considered as an aggregate customer whose demand equals $d_{i_1} + d_{j_1}$. When either i_1 or j_1 appears in a subsequent pair in the list, we must associate this aggregate demand with either customer. For example, suppose the next pair on the list equals (i_1, j_2). What we effectively consider at this point is whether customer j_2 can be added to the route already containing i_1, which can only occur if $d_{i_1} + d_{j_1} + d_{j_2} \leq K$. If this is the case, then customer j_2 is added to the route already containing i_1 and j_1, and the new aggregate demand of this route equals $d_{i_1} + d_{j_1} + d_{j_2}$. Suppose that at the lth pair on the list we encounter the pair of customers (i_l, j_l). Let $D(i_l)$ $(D(j_l))$ denote the aggregate demand of the route containing customer i_l (j_l). If $D(i_l) + D(j_l) \leq K$, then the route containing customer i_l is combined with the route containing customer j_l, and the associated aggregate demand equals $D(i_l) + D(j_l)$; otherwise, we continue to the next pair on the list. After processing the last pair on the list, we will have a set of customer routes, i.e., subsets of customers who will share the same vehicle. For each such customer subset, we then solve a TSP containing the customers in the subset and the DC.

Example 9.7 Consider a vehicle routing problem with 12 customers, in which vehicles have a capacity of 300 units. Customer demands are shown in the table below.

Customer, i	1	2	3	4	5	6	7	8	9	10	11	12
Demand, d_i	120	90	80	130	75	85	95	110	115	100	90	75

The distance matrix is based on the Euclidean distances between the location coordinates shown in the table below, where location 0 corresponds to the DC (the coordinates were generated randomly on a 25 × 25 coordinate plane).

	x-coordinate	y-coordinate
0	0.4	11.2
1	4.0	3.2
2	7.5	7.0
3	2.5	0.6
4	10.5	6.4
5	19.0	23.8
6	19.7	8.2
7	2.9	8.0
8	16.9	9.7
9	14.4	6.8
10	25.0	10.7
11	21.9	10.0
12	21.8	22.2

If we apply the savings method, the pair with the highest savings is $(5, 12)$; because $d_5 + d_{12} = 150 < 300$, this pair may be combined on a route. The next pair on the list is $(10, 11)$; because $d_{10} + d_{11} = 190 < 300$, this pair may be combined on a route. The third pair on the list is $(6, 10)$; because $d_{10} + d_{11} + d_6 = 275 < 300$, customer 6 may be added to the route that contains customer 10, forming a route with customers 6, 10, and 11. The next feasible combination of customers into a route occurs at the 16th pair on the list, $(8, 9)$, which combine to form a route with total demand 225 (Exercise 9.8 asks the reader to compute the savings for each customer pair). Following this, the next feasible combination of customers occurs at the 27th pair on the list, $(1, 3)$, which combine to form a route with a demand of 200. The next pair that leads to a feasible combination is the 28th pair, $(2, 4)$, which combine to create a route with a demand of 220. Only one additional pair on the list leads to a feasible combination of customers: the 52nd pair on the list, $(1, 7)$. Customer 7 may be added to the route that contains customers 1 and 3, to form a route with a total demand of 295. The (five) routes produced by the savings method for this example are shown in the table below. The total distance traveled in the solution equals 180.289.

Route	Customers	Demand	Distance
1	5, 12	150	49.865
2	6, 10, 11	275	50.113
3	8, 9	225	35.108
4	2, 4	220	22.454
5	1, 3, 7	295	22.749

If we use customers 2, 6, 7, 8, and 12 as seed points in the GAP heuristic method and solve the resulting instance of the GAP, we obtain the same routes as shown in the table (this occurs if we use the seed points 1, 4, 8, 11, and 12 as well). If we apply the exact branch-and-price solution method, it turns out that these routes are indeed optimal for this instance of the VRP, and the problem is solved at the root node without the need for branching (at the root node, the optimal LP relaxation solution for the set partitioning formulation is binary). Figure 9.14 shows the resulting optimal routes on the coordinate plane. □

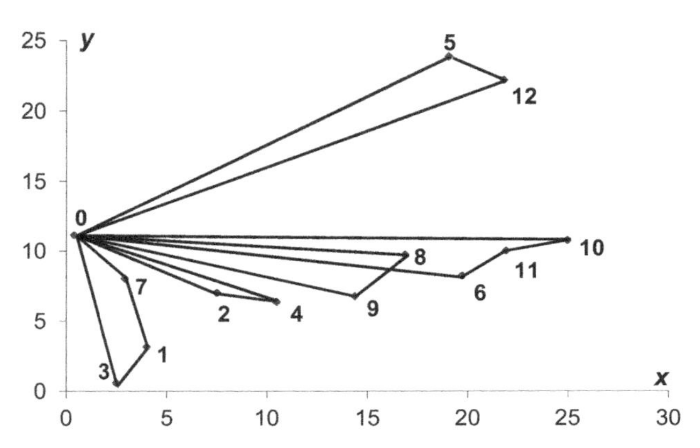

FIGURE 9.14
Optimal routes for Example 9.7.

9.6.4 Additional heuristic methods for the VRP

The heuristic methods we have discussed for the VRP thus far have been cluster-first, route-second methods. In contrast, consider a route-first, cluster-second approach, in which we begin with a TSP solution and then decompose this solution into vehicle routes. Suppose (without loss of generality) that we have n customers, and the TSP tour begins at node 0 (the DC) and visits customers in the sequence $1, 2, \ldots, n$, before returning to the DC. Beasley [16] showed that we can optimally decompose the TSP tour into vehicle routes (note that this does not guarantee an optimal VRP solution, just the best decomposition of a given TSP tour into routes such that the visitation sequence for customers in each route corresponds to the visitation sequence in the TSP solution). To do this, we begin with a directed graph containing nodes 0 through n. Define the arc cost $\hat{c}_{ij}$ as the cost of a route that begins at node 0, visits customers $i+1$ through j (in index order) and then returns to the depot. An arc exists in the network for all node pairs (i, j) with $j > i$ such that $\sum_{l=i+1}^{j} d_l \leq K$ (assuming all vehicles have capacity K). Then the shortest path in this graph gives the lowest cost decomposition of the tour into feasible routes. If arc (i, j) is contained in this shortest path, then a route is created that visits customers $i+1$ through j and returns to the DC. Unfortunately, we cannot say that the optimal decomposition of the optimal TSP solution also gives an optimal solution for the VRP.

In addition to the methods we have discussed thus far, it is straightforward to adapt the heuristic methods we discussed for the TSP in order to provide heuristic solutions for the VRP. Consider, for example the nearest neighbor method. We may start with the nearest customer to the depot and next, after reaching the first customer, move to the nearest feasible neighbor (i.e., the nearest customer whose demand, when added to the route, does not violate the vehicle capacity constraint). The nearest insertion method may be adapted similarly, by applying the nearest feasible insertion. When no feasible insertions exist, the route is completed and a new one is started beginning with the nearest neighbor to the DC that is not part of a previously constructed and completed route. The sweep method has also been applied to the VRP to obtain good heuristic solutions (see Gillett and Miller [56]). In applying the sweep method to the VRP, we keep track of the cumulative demand associated with customers encountered by the sweep arm. As soon as the sweep arm hits a customer that will cause the cumulative demand to exceed the vehicle capacity, we begin a new route starting with this customer, reset the cumulative demand to zero, and end the previous route with the previous customer that touched the sweep arm. Christofides, Mingozzi, and Toth [26] generalized the minimum spanning tree approach to the TSP in order to provide lower bounds on the optimal VRP solution. Their bounding approach starts with a minimum spanning tree such that the DC has degree v for some v such that $|\mathcal{V}| \leq v \leq 2|\mathcal{V}|$, and then adding to this the $|\mathcal{V}|$ least cost edges such that the DC has degree $2|\mathcal{V}|$. Fisher [46] used a different $|\mathcal{V}|$-tree approach that led to improved lower bounds for the VRP.

9.7 Review

Traveling salesman and vehicle routing problems are routinely solved in practice in various contexts. The practicality of these problems along with their interesting structure and complexity have resulted in significant attention by researchers over many years. The scope of this chapter is intended to introduce the reader to these problem classes as well as the kinds of solution approaches that have been applied to them with varying degrees of success. As we have seen, because of the complexity of these problems, research has often focused on effective lower bounding and heuristic techniques. This chapter, however, just scratches the surface of the volume of material available on the TSP and VRP. Like several of the topics covered in this book, sufficient material exists on the TSP and VRP to fill several books. The books by Applegate, Bixby, Chvátal, and Cook [7], Cook [31], Lawler, Lenstra, Rinnooy Kan, and Shmoys [73], Golden, Raghavan, and Wasil [58], and Toth and Vigo [109] consider these problems in much greater depth, and are recommended for researchers who wish to focus on these areas.

Exercises

Ex. 9.1 — Provide an expression for the number of subtour elimination constraints of the form (9.4) as a function of n, the number of cities to be visited.

Ex. 9.2 — Show the validity of inequality (9.8) by starting with the relevant valid subtour elimination constraints of the form (9.4) for the sets T_r, $H \cap T_r$, and $T_r \backslash H$ for $r = 1, \ldots, k$.

Ex. 9.3 — For the graph shown in Figure 9.5, generalize the graph to the case of n nodes, where the greatest arc length equals 2^{n-3}, and derive Equation (9.11) for the worst-case relative performance of the nearest neighbor solution.

Ex. 9.4 — The Hamiltonian tour shown in Figure 9.11 was obtained by duplicating each arc in the MST in Figure 9.10, walking a Eulerian tour, and taking shortcuts to eliminate multiple visits to a node. Show a different Eulerian tour and set of shortcuts that leads to a different Hamiltonian tour.

Ex. 9.5 — Consider a TSP graph with $n+1$ nodes. Characterize the number of options we have for choosing k arcs to remove from a given solution under the k-opt method. In addition, characterize the number of ways in which a solution in which k arcs are removed may be pieced back together.

Ex. 9.6 — For the VRP, given a solution that is not optimal, describe some potential kinds of local neighborhood exchanges and the corresponding local neighborhood definitions.

Ex. 9.7 — Explain how you might use the VRPGAP formulation to find good solutions for instances of the VRP in which splitting customer deliveries among different vehicles is permitted.

Ex. 9.8 — For the problem in Example 9.7, generate the distance matrix, compute the savings values for each customer pair using Equation (9.24), and list these values in nonincreasing order.

Bibliography

[1] K. Aardal. Capacitated facility location: Separation algorithms and computational experience. *Mathematical Programming*, 81:149–175, 1998.

[2] K. Aardal, Y. Pochet, and L.A. Wolsey. Capacitated facility location: Valid inequalities and facets. *Mathematics of Operations Research*, 20(3):562–582, 1995.

[3] P. Afentakis and B. Gavish. Optimal lot-sizing algorithms for complex product structures. *Operations Research*, 34(2):237–249, 1986.

[4] P. Afentakis, B. Gavish, and U. Karmarkar. Computationally efficient optimal solutions to the lot-sizing problem in multistage assembly systems. *Management Science*, 30(2):222–239, 1984.

[5] A. Aggarwal and J.K. Park. Improved algorithms for economic lot-size problems. *Working Paper, Laboratory for Computer Science, Massachusetts Institute of Technology, Cambridge, MA, USA*, 1990.

[6] R. Ahuja, T. Magnanti, and J. Orlin. *Network Flows: Theory, Algorithms, and Applications.* Prentice Hall, Englewood Cliffs, NJ, USA, 1993.

[7] D.L. Applegate, R.E. Bixby, V. Chvátal, and W.J. Cook. *The Traveling Salesman Problem: A Computational Study.* Princeton University Press, Princeton, NJ, USA, 2011.

[8] E. Arkin, D. Joneja, and R. Roundy. Computational complexity of uncapacitated multi-echelon production planning problems. *Operations Research Letters*, 8(2):61–66, 1989.

[9] B.S. Baker. A new proof for the first-fit decreasing bin packing algorithm. *Journal of Algorithms*, 6:49–70, 2002.

[10] A. Balakrishnan and J. Geunes. Production planning with flexible product specifications: An application to specialty steel manufacturing. *Operations Research*, 51(1):94–112, 2003.

[11] E. Balas. The prize collecting traveling salesman problem. *Networks*, 19(6):621–636, 1989.

[12] I. Barany, T.J. Van Roy, and L.A. Wolsey. Uncapacitated lot sizing: The convex hull of solutions. *Mathematical Programming Study*, 22:32–43, 1984.

[13] C. Barnhart, E.L. Johnson, G.L. Nemhauser, M.W.P. Savelsbergh, and P.H. Vance. Column generation for solving huge integer programs. *Operations Research*, 46(3):316–329, 1998.

[14] M.S. Bazaraa, J.J. Jarvis, and H.D. Sherali. *Linear Programming and Network Flows.* John Wiley & Sons, Hoboken, NJ, USA, 4th edition, 2010.

[15] M.S. Bazaraa, H.D. Sherali, and C.M. Shetty. *Nonlinear Programming: Theory and Algorithms.* John Wiley & Sons, Hoboken, NJ, USA, 3rd edition, 2006.

[16] J.E. Beasley. Some applications of the theory of dynamic programming-A review. *Omega*, 11(4):403–408, 1954.

[17] R. Bellman. Route first–cluster second methods for vehicle routing. *Operations Research*, 2(2):275–188, 1983.

[18] J.F. Benders and J.A.E.E. van Nunen. A property of assignment type mixed integer linear programming problems. *Operations Research Letters*, 2(2):47–52, 1983.

[19] J. Bramel and D. Simchi-Levi. *The Logic of Logistics.* Springer, New York, NY, USA, 1997.

[20] L. Buschkühl, F. Sahling, S. Helber, and H. Tempelmeier. Dynamic capacitated lot-sizing problems: A classification and review of solution approaches. *OR Spectrum*, 32:231–261, 2010.

[21] J. Byrka and K. Aardal. An optimal bifactor approximation algorithm for the metric uncapacitated facility location problem. *SIAM Journal on Computing*, 39(6):2212–2231, 2010.

[22] L.M.A. Chan, A. Muriel, Z.-J. Shen, D. Simchi-Levi, and C.-P. Teo. Effective zero-inventory-ordering policies for the single-warehouse multiretailer problem with piecewise linear cost structures. *Management Science*, 48:1446–1460, 2000.

[23] M. Charikar and S. Guha. Improved combinatorial algorithms for facility location and k-median problems. In *Proceedings of the 40th IEEE Symposium on the Foundations of Computer Science*, pages 378–388, 1999.

[24] S. Chopra and P. Meindl. *Supply Chain Management: Strategy, Planning, and Operation.* Pearson, Upper Saddle River, NJ, USA, 5th edition, 2013.

[25] N. Christofides. The traveling salesman problem. In N. Christofides, A. Mingozzi, P. Toth, and C. Sandi, editors, *Combinatorial Optimization.* John Wiley & Sons, New York, NY, USA, 1979.

[26] N. Christofides, A. Mingozzi, and P. Toth. Exact algorithms for the vehicle routing problem, based on spanning tree and shortest path relaxations. *Mathematical Programming*, 20(1):255–282, 1981.

[27] F.A. Chudak and D.B. Shmoys. Improved approximation algorithms for the uncapacitated facility location problem. *SIAM Journal on Computing*, 33(1):1–25, 2003.

[28] F.A. Chudak and D.P. Williamson. Improved approximation algorithms for capacitated facility location problems. *Mathematical Programming Series A*, 102:207–222, 2005.

[29] V. Chvátal. Edmonds polytopes and weakly Hamiltonian graphs. *Mathematical Programming*, 5:29–40, 1973.

[30] G. Clarke and J. Wright. Scheduling of vehicles from a central depot to a number of delivery points. *Operations Research*, 12(4):568–581, 1964.

[31] W.J. Cook. *In Pursuit of the Traveling Salesman: Mathematics at the Limits of Computation.* Princeton University Press, Princeton, NJ, USA, 2011.

[32] G. Cornuejols, G.L. Nemhauser, and L.A. Wolsey. The uncapacitated facility location problem. In P. Mirchandani and R.L. Francis, editors, *Discrete Location Theory*. John Wiley & Sons, New York, NY, USA, 1990.

[33] G.A. Croes. A method for solving traveling-salesman problems. *Operations Research*, 6(6):791–812, 1958.

[34] W.B. Crowston and M. Wagner. Dynamic lot size models for multi-stage assembly systems. *Management Science*, 20(1):14–21, 1973.

[35] W.B. Crowston, M. Wagner, and J.F. Williams. Economic lot size determination in multi-stage assembly systems. *Management Science*, 19(5):791–812, 1973.

[36] G.B. Dantzig. Reminiscences about the origins of linear programming. *Operations Research Letters*, 1(2):43–48, 1982.

[37] I.R. De Farias, Jr., E.L. Johnson, and G.L. Nemhauser. A generalized assignment problem with special ordered sets: a polyhedral approach. *Mathematical Programming*, 89(1):187–203, 2000.

[38] I.R. De Farias, Jr. and G.L. Nemhauser. A family of inequalities for the generalized assignment polytope. *Operations Research Letters*, 29(2):49–55, 2001.

[39] M. Denizel. Minimization of the number of tool magazine setups on automated machines: A Lagrangean decomposition approach. *Operations Research*, 51(2):309–320, 2003.

[40] M. Desrochers, J. Desrosiers, and M. Solomon. A new optimization algorithm for the vehicle routing problem with time windows. *Operations Research*, 40(2):342–354, 1992.

[41] M. Englert, H. Röglin, and B. Vöcking. Worst case and probabilistic analysis of the 2-Opt algorithm for the TSP. In *Proceedings of the Eighteenth Annual ACM-SIAM Symposium on Discrete Algorithms*, pages 1295–1304, 2007.

[42] D. Erlenkotter. A dual-based procedure for uncapacitated facility location. *Operations Research*, 26(6):992–1009, 1978.

[43] A. Federgruen and M. Tzur. A simple forward algorithm to solve general dynamic lot sizing models with n periods in $\mathcal{O}(n \log n)$ or $\mathcal{O}(n)$ time. *Management Science*, 37(8):909–925, 1991.

[44] A. Federgruen and Y.-S. Zheng. Optimal power-of-two replenishment strategies in capacitated general production/distribution networks. *Management Science*, 39(6):710–727, 1993.

[45] M.L. Fisher. The Lagrangian relaxation method for solving integer programming problems. *Management Science*, 27(1):1–18, 1981.

[46] M.L. Fisher. Optimal solution of vehicle routing problems using minimum K-trees. *Operations Research*, 42(4):626–642, 1994.

[47] M.L. Fisher and R. Jaikumar. A generalized assignment heuristic for vehicle routing. *Networks*, 11:109–124, 1981.

[48] M.L. Fisher, R. Jaikumar, and L.N. Van Wassenhove. A multiplier adjustment method for the generalized assignment problem. *Management Science*, 32(9):1095–1103, 1986.

[49] M. Florian and M. Klein. Deterministic production planning with concave costs and capacity constraints. *Management Science*, 18(1):12–20, 1971.

[50] M. Florian, J.K. Lenstra, and A.H.G. Rinnooy Kan. Deterministic production planning: Algorithms and complexity. *Management Science*, 26(7):669–679, 1980.

[51] B.A. Foster and D. Ryan. An integer programming approach to the vehicle scheduling problem. *Operational Research Quarterly*, 27:367–384, 1976.

[52] M.R. Garey, R.L. Graham, and D.S. Johnson. Some $\mathcal{NP}$-complete geometric problems. In *Proceedings of the 8th Annual ACM Symposium on the Theory of Computing*, pages 10–22, 1976.

[53] M.R. Garey, R.L. Graham, D.S. Johnson, and A.C. Yao. Resource constrained scheduling as generalized bin packing. *Journal of Combinatorial Theory, Series A*, 21:257–298, 1976.

[54] M.R. Garey and D.S. Johnson. *Computers and Intractability: A Guide to the Theory of $\mathcal{NP}$-Completeness.* W.H. Freeman & Co., New York, NY, USA, 1979.

[55] A.M. Geoffrion. Lagrangean relaxation for integer programming. *Mathematical Programming Study*, 2:82–114, 1974.

[56] B.E. Gillett and L.R. Miller. A heuristic algorithm for the vehicle-dispatch problem. *Operations Research*, 22(2):340–349, 1974.

[57] M.X. Goemans. Worst-case comparison of valid inequalities for the TSP. *Mathematical Programming*, 69:335–349, 1995.

[58] B.L. Golden, S. Raghavan, and E.A. Wasil, editors. *The Vehicle Routing Problem: Latest Advances and New Challenges.* Springer, New York, NY, USA, 2008.

[59] E.S. Gottlieb and M.R. Rao. The generalized assignment problem: Valid inequalities and facets. *Mathematical Programming*, 46(1–3):31–52, 1990.

[60] M. Grötschel and M.W. Padberg. On the symmetric travelling salesman problem. I. Inequalities. *Mathematical Programming*, 16:265–280, 1979.

[61] M. Grötschel and W.R. Pulleyblank. Clique tree inequalities and the symmetric travelling salesman problem. *Mathematics of Operations Research*, 11(4):537–569, 1986.

[62] S. Guha and S. Khuller. Greedy strikes back: Improved facility location algorithms. *Journal of Algorithms*, 31:228–248, 1999.

[63] M. Guignard and S. Kim. A strong Lagrangian relaxation for capacitated plant location problems. *Working Paper No. 56, Department of Statistics, University of Pennsylvania*, 1983.

[64] M. Haimovich and A.H.G. Rinnooy Kan. Bounds and heuristics for capacitated routing problems. *Mathematics of Operations Research*, 10(4):527–542, 1985.

[65] F.W. Harris. How many parts to make at once. *Factory, The Magazine of Management*, 10(2):135–136, 152, 1913.

[66] M. Held, P. Wolfe, and H.P. Crowder. Validation of subgradient optimization. *Mathematical Programming*, 6:62–88, 1974.

[67] P.L. Jackson, W.L. Maxwell, and J.A. Muckstadt. Determining optimal reorder intervals in capacitated production-distribution systems. *Management Science*, 34(8):938–958, 1988.

[68] R.M. Karp. Reducibility among combinatorial problems. In R.E. Miller and J.W. Thatcher, editors, *Complexity of Computer Computations*, pages 85–103. Plenum, New York, NY, USA, 1972.

[69] H. Kellerer, U. Pferschy, and D. Pisinger. *Knapsack Problems*. Springer-Verlag, Berlin, Germany, 2004.

[70] M. Korupolu, C. Plaxton, and R. Rajaraman. Analysis of a local search heuristic for facility location problems. *Journal of Algorithms*, 37:146–188, 2000.

[71] A.A. Kuehn and M.J. Hamburger. A heuristic program for locating warehouses. *Management Science*, 9:643–666, 1963.

[72] E.L. Lawler. Fast approximation algorithms for knapsack problems. *Mathematics of Operations Research*, 4(4):339–356, 1979.

[73] E.L. Lawler, J.K. Lenstra, A.H.G. Rinnooy Kan, and D.B. Shmoys. *The Traveling Salesman Problem: A Guided Tour of Combinatorial Optimization*. John Wiley & Sons, Hoboken, NJ, USA, 1985.

[74] J.M.Y. Leung and T.L. Magnanti. Valid inequalities and facets of the capacitated plant location problem. *Mathematical Programming*, 44:271–291, 1989.

[75] J.M.Y. Leung, T.L. Magnanti, and R. Vachani. Facets and algorithms for capacitated lot sizing. *Mathematical Programming*, 45:331–359, 1989.

[76] R. Levi, R. Roundy, and D. Shmoys. Primal-dual algorithms for deterministic inventory problems. *Mathematics of Operations Research*, 31(2):267–284, 2006.

[77] R. Levi, R. Roundy, D. Shmoys, and M. Sviridenko. A constant approximation algorithm for the one-warehouse multiretailer problem. *Management Science*, 54(4):763–776, 2008.

[78] R. Levi, D.B. Shmoys, and C. Swamy. Lp-based approximation algorithms for capacitated facility location. In *Integer Programming and Combinatorial Optimization*, volume 3064 of *Lecture Notes in Computer Science*, pages 206–218. Springer, Berlin, Germany, 2004.

[79] S. Li. A 1.488 approximation algorithm for the uncapacitated facility location problem. *Information and Computation*, 222:45–58, 2012.

[80] S. Lin and B.W. Kernighan. An effective heuristic algorithm for the traveling-salesman problem. *Operations Research*, 21(2):498–516, 1973.

[81] S.F. Love. A facilities in series inventory model with nested schedules. *Management Science*, 18(5):327–338, 1972.

[82] M. Mahdian, Y. Ye, and J. Zhang. Approximation algorithms for metric facility location problems. *SIAM Journal on Computing*, 36(2):411–432, 2006.

[83] S. Martello and P. Toth. An algorithm for the generalized assignment problem. In J.P. Brans, editor, *Operational Research*. IFORS, North-Holland, Amsterdam, The Netherlands, 1981.

[84] S. Martello and P. Toth. *Knapsack Problems: Algorithms and Computer Implementations*. Wiley, Chichester, West Sussex, UK, 1990.

[85] W.L. Maxwell and J.A. Muckstadt. Establishing consistent and realistic reorder intervals in production-distribution systems. *Operations Research*, 33(6):1316–1341, 1985.

[86] R.M. Nauss. An improved algorithm for the capacitated facility location problem. *Journal of the Operational Research Society*, 29(12):1195–1201, 1978.

[87] G.L. Nemhauser and L.A. Wolsey. *Integer and Combinatorial Optimization*. John Wiley & Sons, New York, NY, USA, 1988.

[88] G.L. Nemhauser, L.A. Wolsey, and M.L. Fisher. An analysis of approximations for maximizing submodular set functions–I. *Mathematical Programming*, 14(1):265–294, 1978.

[89] M.W. Padberg, T.J. Van Roy, and L.A. Wolsey. Valid linear inequalities for fixed charge problems. *Operations Research*, 33(4):842–861, 1985.

[90] M. Pál, É Tardos, and T. Wexler. Facility location with nonuniform hard capacities. In *Proceedings of the 41st IEEE Symposium on the Foundations of Computing*, pages 329–338, 2001.

[91] C.H. Papadimitriou. The Euclidean traveling salesman problem is $\mathcal{NP}$-Complete. *Theoretical Computer Science*, 4:237–244, 1977.

[92] C.H. Papadimitriou, A.A. Schäffer, and M. Yannakakis. On the complexity of local search. In *Proceedings of the twenty-second annual ACM symposium on Theory of computing*, pages 438–445, 1990.

[93] Y. Pochet. Valid inequalities and separation for capacitated economic lot sizing. *Operations Research Letters*, 7(3):109–115, 1988.

[94] B.T. Poljak. A general method of solving extremum problems. *Soviet Mathematics Doklady*, 8:593–597, 1967.

[95] H.E. Romeijn and D. Romero Morales. A class of greedy algorithms for the generalized assignment problem. *Discrete Applied Mathematics*, 103:209–235, 2000.

[96] D.J. Rosenkrantz, R.E. Stearns, and P.M. Lewis II. An analysis of several heuristics for the traveling salesman problem. *SIAM Journal on Computing*, 6(3):563–581, 1977.

[97] G.T. Ross and R.M. Soland. A branch and bound algorithm for the generalized assignment problem. *Mathematical Programming*, 8(1):91–103, 1975.

[98] R. Roundy. 98%-effective integer-ratio lot-sizing for one-warehouse multi-retailer systems. *Management Science*, 31(11):1416–1430, 1985.

[99] R. Roundy. A 98%-effective lot-sizing rule for a multi-product, multi-stage production/inventory system. *Mathematics of Operations Research*, 11(4):699–727, 1986.

[100] R. Roundy. Rounding off to powers of two in continuous relaxations of capacitated lot sizing problems. *Management Science*, 35(12):1433–1442, 1989.

[101] R. Roundy. Efficient, effective lot sizing for multistage production systems. *Operations Research*, 41(2):371–385, 1993.

[102] D.M. Ryan and B.A. Foster. An Integer Programming Approach to Scheduling. In A. Wren, editor, *Computer Scheduling of Public Transport Urban Passenger Vehicle and Crew Scheduling*, pages 269–280. North-Holland, Amsterdam, The Netherlands, 1981.

[103] H.M. Salkin and K. Mathur. *Foundations of Integer Programming*. North-Holland, New York, NY, USA, 1989.

[104] M.W.P. Savelsbergh. A branch-and-price algorithm for the generalized assignment problem. *Operations Research*, 45(6):831–841, 1997.

[105] Z.-J. Shen, J. Shu, D. Simchi-Levi, C.-P. Teo, and J. Zhang. Approximation algorithms for general one-warehouse multi-retailer systems. *Naval Research Logistics*, 56(7):642–658, 1009.

[106] D.B. Shmoys, É Tardos, and K. Aardal. Approximation algorithms for facility location problems. In *Proceedings of the Twenty-Ninth Annual ACM Symposium on Theory of Computing*. ACM, 1997.

[107] D. Simchi-Levi. New worst-case results for the bin-packing problem. *Naval Research Logistics*, 41:579–585, 1994.

[108] M. Sviridenko. An improved approximation algorithm for the metric uncapacitated facility location problem. In W.J. Cook and A.S. Schulz, editors, *Integer Programming and Combinatorial Optimization: 9th International IPCO Conference, Cambridge, MA*, volume 2337 of *Lecture Notes in Computer Science*, pages 240–257. Springer, New York, NY, USA, 2002.

[109] P. Toth and D. Vigo, editors. *The Vehicle Routing Problem*. SIAM, Philadelphia, PA, USA, 2001.

[110] C.P.M. van Hoesel. *Algorithms for Single-Item Lot-Sizing Problems (Ph.D. Dissertation)*. Erasmus University Rotterdam, Rotterdam, The Netherlands, 1991.

[111] C.P.M. van Hoesel and A.P.M. Wagelmans. An $O(T^3)$ algorithm for the economic lot-sizing problem with constant capacities. *Management Science*, 42(1):142–150, 1996.

[112] C.P.M. van Hoesel and A.P.M. Wagelmans. Fully polynomial approximation schemes for single-item capacitated economic lot-sizing problems. *Mathematics of Operations Research*, 26(2):339–357, 2001.

[113] T.J. Van Roy. A cross decomposition algorithm for capacitated facility location. *Operations Research*, 34(1):145–163, 1986.

[114] P.H. Vance, C. Barnhart, E.L. Johnson, and G.L. Nemhauser. Solving binary cutting stock problems by column generation and branch-and-bound. *Computational Optimization and Applications*, 3(2):111–130, 1994.

[115] A.P.M. Wagelmans, C.P.M. van Hoesel, and A. Kolen. An $\mathcal{O}(n \log n)$ algorithm that runs in linear time in the Wagner-Whitin case. *Operations Research*, 40(Supp. 1):S145–S156, 1992.

[116] H. Wagner and T. Whitin. Dynamic version of the economic lot size model. *Management Science*, 5:89–96, 1958.

[117] W.I. Zangwill. A deterministic multi-period production scheduling model with backlogging. *Management Science*, 13(1):105–119, 1966.

[118] W.I. Zangwill. A backlogging model and a multi-echelon model of a dynamic economic lot size production system–A network approach. *Management Science*, 15(9):506–527, 1969.

[119] J. Zhang, B. Chen, and Y. Ye. A Multiexchange Local Search Algorithm for the Capacitated Facility Location Problem. *Mathematics of Operations Research*, 30(2):389–403, 2005.

[120] P.H. Zipkin. *Foundations of Inventory Management*. McGraw-Hill, Boston, MA, USA, 2000.

Index

Asymptotic Optimality, 4, 15
Asymptotically Optimal Heuristic, 4
 0-1 Knapsack Problem, 15–16
 Generalized Assignment Problem, 51–53

Bill-of-Materials, 95
Bin Packing Problem, 27–29, 156
 Definition, 27
 Heuristics
  Best-Fit, 28
  Best-Fit Decreasing, 28
  First-Fit, 28
  First-Fit Decreasing, 28
Branch-and-Bound Algorithm, 5, 10, 20, 27, 34–36, 47, 49, 102, 103, 105, 160, 173
Branch-and-Price, 5
 Branching Strategy, 35
 Generalized Assignment Problem, 47–49
 Set Partitioning, 33–36
 Vehicle Routing Problem, 183–185

Column Generation, 5, 30
 Generalized Assignment Problem, 48
 Pricing Problem, 31
 Set Partitioning, 29–33
 Vehicle Routing Problem, 184–185
Complexity, 4
 0-1 Knapsack Problem, 10, 11
  Fully Polynomial Time Approximation Scheme, 17, 19
 Capacitated Lot Sizing Problem, 75
  Fully Polynomial Time Approximation Scheme, 85
 Facility Location Problem, 140, 142–143
 Generalized Assignment Problem, 39
 $\mathcal{NP}$-Completeness, 10
 $\mathcal{NP}$-Hardness, 9, 10
 Pseudopolynomial Time Algorithm, 10
 Serial System Lot Sizing, 100
 Set Partitioning Problem, 27
 Traveling Salesman Problem, 169
Constant-Factor Approximation, 4, 149
 Joint Replenishment Problem, 111
 Metric Capacitated Facility Location Problem, 162–165
 Metric Uncapacitated Facility Location Problem, 148–154
 One-Warehouse Multi-Retailer System, 105–111
Convex Hull, 20–22, 42, 43, 54, 86–89, 92, 93, 157, 160, 161, 172, 175
Cover Inequalities
 0-1 Knapsack Problem, 21–22
 Capacitated Facility Location Problem, 160–162
 Capacitated Lot Sizing Problem, 90–92

Dual-Ascent Algorithm, 143–148
Dynamic Programming
 Capacitated Lot Sizing Problem, 77–85
  Formulation, 78
 Knapsack Problem, 9
 Serial System, 99–100
 Uncapacitated Lot Sizing Problem, 61–62
  Formulation, 61
  Formulation with Backordering, 71
 Vehicle Routing Pricing Problem, 184

Echelon Holding Cost, 101, 102, 114
Echelon Inventory, 101–104, 114

Economic Order Quantity (EOQ), 57, 96, 112, 114
  Multistage EOQ, 112–134
    Formulation, 116
    Solution Method, 117–122
Eulerian Tour, 180, 181

Facets
  0-1 Knapsack Problem, 21
  Capacitated Facility Location Problem, 160, 161
  Capacitated Lot Sizing Problem, 88, 89, 92
  Definition, 20
  Generalized Assignment Problem, 54
  Traveling Salesman Problem, 172, 175
  Uncapacitated Lot Sizing Problem, 87
Facility Location Problem, 5, 75, 137–165
  All-or-Nothing Demand, 140
  Capacitated
    Constant-Factor Approximation, 162–165
    Lagrangian Relaxation, 154–157
    Valid Inequalities, 157–162
  Complexity, 143
  Cost-Minimization Version
    Formulation, 141
  Metric, 5
    Constant-Factor Approximation, 148–154
    Uncapacitated, 148–154
  Profit-Maximizing Formulation, 139
  Single-Facility Version, 139
  Single-Sourcing, 140, 142
  Uncapacitated
    Dual-Ascent Algorithm, 143–148
Fully Polynomial Time Approximation Scheme (FPTAS), 4, 112
  0-1 Knapsack Problem, 17–19
  Capacitated Lot Sizing, 80–85

Generalized Assignment Problem, 5, 142, 185–187
  Asymptotically Optimal Heuristic, 51–53
  Branch-and-Price, 47–49
  Complexity, 39
  Formulation, 40
  Lagrangian Relaxation, 46–47
  Set Partitioning Formulation, 48
    Branching Strategy, 49
    Pricing Problem, 48
Greedy Algorithm, 12
  Bin Packing Problem, 28
  Generalized Assignment Problem, 49–54
  Traveling Salesman Problem, 175

Hamiltonian Circuit, 169, 185
  Definition, 32

Joint Replenishment Problem, 128–130
  Dynamic Demand, 111
  Linear Relaxation, 129–130
  Power-of-Two Policy
    Effectiveness, 129
  Power-of-Two Solution, 130

Karush Kuhn Tucker (KKT) Conditions, 116
Knapsack Problem, 4, 25, 39
  0-1 Knapsack Problem, 8, 33, 49, 75, 140, 156
    Asymptotically Optimal Heuristic, 15–16
    Complexity, 10
    Cover Inequalities, 21–22
    Fully Polynomial Time Approximation Scheme, 17–19
    Linear Programming Relaxation, 11, 19
    Linear Relaxation, 10
    $\mathcal{NP}$-Completeness, 11
    Valid Inequalities, 19–22
  Continuous Knapsack Problem, 12, 139, 156
    Greedy Algorithm, 12

Lagrangian Relaxation, 5
  Capacitated Facility Location Problem, 154–157
  Formulation, 42
  General Technique, 41–45
  Generalized Assignment Problem, 46–47

Integrality Property, 44, 154
Lagrangian Dual Problem, 43, 105
Multistage Assembly System Lot Sizing Problem, 102
Multistage System Lot Sizing Problem, 104
Subgradient Optimization, 44, 45, 103, 105
Uncapacitated Lot Sizing Problem, 63–65
Linear Relaxation, 10
Joint Replenishment Problem, 129–130
Linear Programming Relaxation, 11
Capacitated Facility Location, 164–165
Facility Location Problem, 144
Metric Uncapacitated Facility Location, 149
One-Warehouse Multi-Retailer Problem, 107–110
Set Partitioning Problem, 29, 184
Multistage EOQ Problem, 116–122
Solution Method, 117–122
Lot Sizing Problem, 4
Assembly System, 101–103
Capacitated, 73–93, 141
Complexity, 75
Fixed-Charge Network Flow Interpretation, 76–77
Formulation, 74
Fully Polynomial Time Approximation Scheme, 80–85
Valid Inequalities, 85–92
Equal-Capacity Case, 78–80
General System, 103–105
Joint Replenishment Problem, 111
One-Warehouse Multi-Retailer System, 105–111
Problem Formulation, 106
Regeneration Interval, 61, 71, 73, 76–79
Serial System, 97–100
Uncapacitated, 57–72, 103, 141
Backordering, 69–71
Fixed-Charge Network Flow Interpretation, 59–61
Lagrangian Relaxation, 63–65
$\mathcal{O}(T \log T)$ Algorithm, 65–69
Problem Formulation, 59
Shortest Path Solution, 62, 71
Tight Formulation, 62–63
Zero-Inventory Ordering Property, 60

Mixed Integer Optimization, 3
MRP Systems, 58
Multistage Inventory Systems, 5, 95–112
Assembly System, 95, 101–103
Problem Formulation, 102
Distribution System, 95
Dynamic Demand, 97–112
General System, 95, 103–105
Joint Replenishment Problem, 128–130
Multistage EOQ, 112–134
Formulation, 116
Power-of-Two Policy, 122–128
Solution Method, 117–122
One-Warehouse Multi-Retailer System, 130–134
Solution Method, 131–133
Serial System, 95, 97–100
Complexity, 100
Problem Formulation, 98

Nested Policy, 100, 102, 104, 112, 114, 115, 124, 125, 129, 130

One-Warehouse Multi-Retailer System, 130–134
Dynamic Demand, 105–111
Problem Formulation, 131
Solution Method, 131–133
Power-of-Two Policy Solution, 133

Planning
Operations, 1, 2
Strategic, 1, 2
Tactical, 2
Power-of-Two Policy, 114, 115, 124–126, 130, 131, 133–135
Effectiveness, 125–128, 133

Regeneration Interval, 61, 71, 73, 76–79

Set Covering Problem, 25, 38, 143
Set Packing Problem, 25, 37

Set Partitioning Problem, 5, 25, 48, 184
  Branch-and-Price, 33–36
  Column Generation, 29–33
  Complexity, 27
  Formulation, 27
  Linear Programming Relaxation, 29
Spanning Tree, 60, 179
Stationary Policy, 113–115, 125, 130, 131
Subset Sum Problem, 11, 76

Total Unimodularity, 46
Traveling Salesman Problem, 32, 167–182
  2-Opt Exchange, 182
  Asymmetric, 168
  Comb Inequalities, 173–175
  Complexity, 169
  Facets, 172, 175
  k-Opt Exchange, 182
  Local Improvement Heuristics, 182
  Minimum Spanning Tree Heuristic, 179–181, 191
  Nearest Neighbor Heuristic, 175–177, 191
  Prize-Collecting, 32, 184
  Subtour Elimination Constraints, 171–173
  Subtours, 171
  Sweep Heuristic, 177–179, 191
  Symmetric, 167, 168

Valid Inequalities, 5
  0-1 Knapsack Problem, 19–22
  Capacitated Lot Sizing Problem, 85–92
  Facility Location Problem, 157–162
Vehicle Routing Problem, 5, 168, 182–191
  Branch-and-Price Approach, 183–185
    Pricing Problem, 184
  Clarke-Wright Savings Heuristic, 187–190
  Cluster-First Route-Second Heuristic, 183, 187
  Generalized Assignment Problem Heuristic, 185–187
  Route-First Cluster-Second Heuristic, 183, 190

Zero-Inventory Ordering Property, 60, 73, 99, 102–104, 113

For Product Safety Concerns and Information please contact our EU representative GPSR@taylorandfrancis.com Taylor & Francis Verlag GmbH, Kaufingerstraße 24, 80331 München, Germany

Printed and bound by CPI Group (UK) Ltd, Croydon, CR0 4YY

08/05/2025

01864442-0001